AF882230

Living Waters
Pulse of the Planet

Living Waters
Pulse of the Planet

Edited by
ROMOLA BUTALIA

Ink Impact
Publishing

First Edition: Delhi, ***2025***

© ROMOLA BUTALIA

Introduction copyright © Romola Butalia 2025
Copyright for the individual essays vests with their respective authors.

All Rights Reserved. No part of this book may be reproduced in any form, by photocopy or by any other electronic means, without permission in writing from the author and publisher.

Also available at:
MOTILAL BANARSIDASS PUBLISHING HOUSE

- 4637/20, F-2, Hari Sadan, First Floor, Ansari Road, Darya Ganj Delhi–110 002 *Tel.*: +91-11-4559 7999
- 1st Floor, 1/2 Rathnavilas Road, Basvanagudi, Next to Pai Vista Convention Hall, South End Circle, Bengaluru–560004 (India) *Tel.*: +91-80-41275911

ISBN: 978-93-7100-299-8 (Hardcover)
ISBN: 978-93-7100-728-3 (Paperback)

Printed & Published by
Rajendra P Jain *for*
Motilal Banarsidass Publishing House
A-44, Naraina Industrial Area, Phase I, New Delhi–110 028 (INDIA)
Email: naraina@mlbd.in Website: www.mlbd.in
Tel.: (+91-11) 4751 1592

Contents

Vasudhaiva Kutumbakam

An Introduction by Romola Butalia

At the entrance of the Parliament of India is inscribed, *Vasudhaiva Kutumbakam*,[1] the Earth is one family. It is reflective of spiritual vision motivating practical action. Enshrined in this are inclusive perspectives, detachment from personal gain and striving towards higher values.

Scientists, environmentalists, writers, those deeply engaged in working on water in different sectors, have contributed essays to this volume recognising that we can no longer work silently in silos. This is a time for cross-sectoral cross-cultural dialogues, to understand and generate awareness about the severe crisis unfolding as we hurtle towards the Sixth Anthropocene extinction. There is an immense need for value-driven, community-focused conversations to co-create a new and resilient paradigm as we face an era of systemic collapse.

Accelerating glacial melt, rivers in spate and the ensuing floods; dams that prevent the natural flow of water; springs that are drying and aquifers that are being depleted; changing monsoon and weather patterns; deforestation, loss of wetlands and biodiversity; plastic pollution, lack of waste management and sanitation with its effect on water quality; migration, tourism

1. https://www.academia.edu/125641365/VASUDHAIVA_KUTUMBAKAM_AN_INTRODUCTION

beyond capacity, loss of indigenous cultures and traditional wisdom; water required for sustainable agriculture and industry; pollution of oceans, loss of coral reefs and marine life, coastal flooding, loss of mangroves; this unending list reflects the many forces impacting the hydrological cycle and the planet. This book begins a dialogue about them.

Dr Vasudha Pant, Chairperson of the Almora Literature Festival (ALF) sought to highlight the multitude of water-related problems, in the introductory sessions at ALF '24 and '25. From my regular conversations with her, and with several scientists and water crusaders, the idea emerged to curate a book of essays covering diverse forms and aspects of water to make the science widely accessible. Any conversation has to involve creative thinkers, philosophers, writers; and they joined this dialogue. *Living Waters – Pulse of the Planet* is the organic growth of an idea, which will hopefully continue to evolve, reaching more people. Together we can implement changes in healing the earth and waters and connected systems, by acknowledging the current trajectory, and course correcting towards a future we must collectively envision.

MLBD Publishing House enthusiastically embraced the idea of a book to generate awareness and deep engagement on the subject. Atika Jain became the person to head the new imprint, Ink Impact Publishing, that will publish books on ecology and environment; gender; and their core speciality, spirituality. Conversations on ecology, climate change and our current reality must be in the forefront of our consciousness; community engagement needs to be spearheaded. We have to acknowledge the extent of disasters that we will be compelled to cope with.

In presenting these essays, there was no obvious and logical sequential order, so I will introduce the authors and their essays in this collaborative effort, in order of the chapters ahead.

The celebrated poet, **Sudeep Sen** has written a book of poems titled *Anthropocene*. Dr. Anjan Ray, who shares my distress at the unfolding reality and believes in networks to create change, introduced me to Sudeep and his poetry. Sudeep was at the time at a poet's residency at Hyderabad, and the opening chapter of this book is a selection of poems he wrote for us there.

There is a cloud, just a solitary cloud
  wafting perilously —
But it is too far in the distance
  for any real hope — for rain.

These words remain as a refrain for me, as I look at the Himalayan skies where I live, unconsciously seeking clouds, rain and hope.

Living in Kumaon Himalaya, I have been observing the increasing water crisis at the ground level, including the drying of springs, often due to neglect, disregard of sustainable practices, disruption of water channels, ground extraction and pollution of ground water. NITI Aayog in a 2018 report[1] states that there are "five million springs across India, of which nearly 3 million are in the Indian Himalayan Region". An estimated 200 million people depend on spring water in India. The report notes, "Springs also serve the important hydrological function of sustaining winter and dry season stream flows in non-glaciated catchments, helping sustain riverine ecosystems and biodiversity across the Himalayan river basins". NITI Aayog acknowledges that nearly half of the perennial springs in the Indian Himalaya have dried or become seasonal and many villages are facing an acute shortage of water for drinking and domestic use.

1. https://dolr.gov.in/springshed/

My interest in the revival of springs, protection of aquifers and water resources led me to several people and organisations working in this field. **Dr. Himanshu Kulkarni** is a Founder Trustee, former Executive Director and currently Scientist (Emeritus) with ACWADAM, apart from being associated with three leading academic institutions. He has supervised, guided and mentored students from many institutions and disciplines on groundwater and continues to work on bringing a strong 'aquifer' based pedagogy to water management and governance in India. Yet, he makes himself accessible, makes time to explain the smallest detail. His conscientiousness and commitment struck me deeply. During this time he has been seriously ill, but he did not miss a deadline, and reached out at every stage to offer his expertise and help to this project. He writes, "Aquifer management holds the key to resolving India's current groundwater crisis because of the sudden shift to overextraction of groundwater (seven decades) from a long legacy of systematic management through sustainable approaches to extraction (several thousands of years)".

International Centre for Integrated Mountain Development (ICIMOD)[1] has been doing remarkable work in bringing the countries of the Third Pole together, focusing on the problems that are common to the region, by sharing information, knowledge and solutions with countries in the Hindu Kush Himalaya, creating a network that looks for solutions to problems facing the Himalaya rather than staying confined within political and national boundaries, which is a short-sighted approach to the climate crisis.

I reached out to **Sanjeev Bhuchar** of ICIMOD, in Kathmandu, Nepal, who plays a pivotal role in efforts to revive and sustainably manage springs across the Hindu Kush

1. https://www.icimod.org

Himalaya (HKH) region. His essay talks about the urgent need for springshed management. He writes, "The challenge is ensuring these efforts spread across the region. Governments, researchers and communities need to work together to protect springs. This means creating policies that treat springs as critical water resources, investing in spring conservation, and promoting practices that recharge aquifers".

The United Nations proclaimed 2025 as the International Year of Glaciers' Preservation,[1] with the intent to raise global awareness about the critical role of glaciers, snow and ice in the hydrological cycle and climate system, together with highlighting the immense impact of changes in the Earth's cryosphere. The depletion of the world's water towers and rapid glacial melt, with glaciers providing freshwater to 2 billion people globally[2] will have unimaginable cascading impacts.

Dr. Arun Bhakta Shrestha of Nepal, played a lead role in the successful Glacial Lake Outburst Flood (GLOF) risk mitigation at Tsho Rolpa glacial lake and is Senior Advisor – Reducing Climate and Environmental Risks at ICIMOD. In his essay, he writes that glaciers are the lifeline of the Himalaya, explaining about snow cover and permafrost and their crucial role in sustaining water resources, ecosystems and livelihoods. "As climate change accelerates, these frozen reservoirs are shrinking, posing significant threats to water security, agriculture, energy production and increasing disaster risks. By examining the science behind the HKH cryosphere, the vulnerabilities as well as the global significance, we highlight the urgent need for action to protect and adapt to a changing cryosphere in a warming world".

1. https://www.un-glaciers.org/en
2. https://www.unesco.org/en/articles/unesco-and-wmo-launch-international-year-glaciers-preservation-2025

Glaciers hold profound cultural and spiritual significance, often considered sacred spaces. Anyone who has watched the receding Gangotri glacier over the past decades, feels the mounting pain of the sheer abuse of Ganga that we are now witnessing. I am singling out the Gangotri glacier as a symbol of all glaciers, because in sheer numbers the people who hold the Ganga sacred, and the emotional, religious, mythological and spiritual associations that are genetically embedded in us, make it a fitting symbol of the plight of glaciers and rivers. Ganga is not merely a river, she is the Mother Goddess, who descended to Earth after thousands of years of *tapasya* or austerities undertaken in seeking her; she was released from Bramha's *kamandal* or water pot, and fell through Mahadeva's jatas or matted locks, so the Earth would be able to bear her weight.

Siddharth Agarwal's essay is written from having walked along the river Ganga from Gangasagar to Gangotri in a six-month journey covering 3,000 km. Having watched the documentary, *Moving Upstream: Ganga,*[1] I invited him to share his experiences of six months of rigour, deep connections and the lifetime of learning it brought. Talking of the construction of the controversial highway to the Char Dhams in Uttarakhand, he remarks: "The residents whom we met along the way, and who powered our journey with their generosity, reminded us that the Char Dham pilgrimage is considered complete only when it is done on foot".

This slowing down is imperative to all climate solutions, and there is a global movement advocating it. **Alpha Lo** is a water researcher, writer and podcaster, who runs Climate Water Project in US, bringing people together in the regenerative water field. He talks of how to restore the water cycle and how that helps with

1. https://www.youtube.com/watch?v=stG5BbMSCkE

hydrating the earth and soil, replenishing groundwater, restoring rains in drought areas, lessening flooding, and slowing down climate change. "It is time for a new and yet very ancient song, the slow water song. The earth's water has its rhythm, its natural speed. Millions of years of evolution has guided the terrestrial environment and atmosphere to harmonise its water dance. And this water dance happens at a slow tempo, a slower vibration than our out-of-control modern speed obsessed culture".

Ali Bin Shahid is a systems architect and regenerative designer, based in Islamabad, Pakistan. Committed to fostering a more just, equitable and sustainable world, his work emphasises quantifying nature's rhythms and integrating them with engineering principles to develop nature-based climate repair strategies, contributing to the restoration of hydrological cycles and the resilience of ecosystems. "When water moves too quickly across the landscape, rushing off rooftops, racing down streets, flooding through degraded soil, it becomes destructive. It carries away topsoil, causes erosion and disappears before it creates benefit. When water moves slowly, soaking into healthy soil, flowing gently through vegetation, pooling temporarily in natural depressions, it becomes the foundation of life". He shares 3 universal principles and five laws, giving examples of how they were applied successfully, and a practical playbook to start your own rain project.

Dr. Vasudha Pant is an environmentalist and agricultural scientist, who is deeply conscious of social context and seeks to bring the past and the future on one platform, wanting to convey the wisdom and knowledge of the past to the generations to whom the baton will be passed on. Her essay titled *Stepwells and Ancient Practices,* draws attention to ancient civilisations, traditional water sources and hydraulics and the stepwells of India. About

Uttarakhand, she writes that it is "not just home to towering peaks and sacred shrines but also to a fascinating network of traditional water management systems known as naulas and dharas. These structures, deeply integrated with the lives of hill communities, serve as vital lifelines, supplying water in a region where springs and small streams are the primary sources of sustenance".

Dr Suman Sahai has had a distinguished career in Genetics and is founder Chairperson of Gene Campaign, a leading research and advocacy organisation, working on food, nutrition and rural livelihoods. Her essay, *Adapting Agriculture to Diminishing Water Availability* talks about how we have lost traditional "wise conservation policies and instead adopted an expansive utilisation of water, increasing supply as demand increased, without paying any attention to managing demand". She says, "We have been brutally wasteful with groundwater with gushing tube wells running for hours, irrespective of requirement". Recognising that one of the most debilitating impacts of climate change is scarcity of water, Gene Campaign started work on making agriculture more resilient in the Himalayan region.

Dr Anjan Ray has worked in pharmaceuticals, oil and gas, energy efficiency, biofuels, biotechnology, sustainable infrastructure and supply chains, climate change, policy formulation and venture capital among other diverse fields, looking at his work through the prism of sustainability. Somewhere in the labyrinth of environmental conversations, we connected because of deeply felt concerns and a shared value. I would seek his opinion because he is equally comfortable with ideating as he is with data, analysis and application. I carry the firm conviction that any genuine environmentally concerned person is both immensely helpful and communicative because he values shared knowledge, resources and a commitment towards giving.

In his essay, he analyses the impact of industry on water, and concludes with, "If we do not curtail consumption, and become frugal in our use of common resources, if we do not collectively hold industries and governments accountable every time and at every level, future generations will suffer the consequences of our irresponsibility".

Hridayesh Joshi is a journalist who takes the time to delve into what he reports on. I have known him to report objectively, while feeling deeply. When he was reporting on the Kedarnath tragedy of 2013, so many images remained entrenched. "When I reached there after the tragedy, there was deathly silence everywhere. Dead bodies were being found behind mountain rocks, and the stories related would simply make one cry". His book, *Rage of the River* is a translation of the original written by him, *Tum Chup Kyon Rahe Kedar?* Hridayesh can be relied on to say it as he sees it, without making it palatable. His integrity and his passion make him a rare journalist in these times when journalism has changed character and focus. As an award-winning journalist, after years in mainstream journalism, almost inevitably, he is now an independent journalist, because it is virtually impossible to believe you need to tell the truth about what you see, and hope to have the comfort of a pay-cheque. He was one of the first people with whom I discussed the concept for this book, and his was one of the first essays to come in. I had wanted him to write a journalist's story on water.

I first heard **Sunita Narain** speaking in the mid-nineties, when sustainability was disregarded as an eccentricity by mainstream developmental economics. Her impassioned but mature speech belied her size and age at that time, and she impressed me by her deep commitment, and the confidence with which she countered the heady days of liberalisation and the mantra of unending growth, with a quiet balance. As sustainable growth, ethical business, frugality, traditional wisdom were relegated to old-fashioned regressive thinking, her continual attempts to generate environmental and ecological awareness among journalists by organising conferences gnawed at the conscience of the few, who recognised that the inevitable fallout of the Western model of growth would create devastation. I have not met her for many years now, but there was no way I could do a book on water without acknowledging that she inspired me with the depth of her efforts. Despite her schedule and commitments, she wrote this chapter.

Varsha Singh is an independent journalist who covers environment, climate change and health issues. Journalists are trained to tell a story from the angle of human impact. In recent times, it is done through an intrusive voyeurism, a camera trained too close to tragedy, with full intent to manipulate the emotions of the reader/viewer. The sensitivity to tell it through the lens of distance and a forced impartiality while each encounter remains in your mind not as a story but as an experience, is what journalism is about. Only those with the courage of their convictions are not afraid to take a stand. Varsha writes about the impact of dams on rivers and people. Anyone who has watched the dislocation caused by dams, the human cost of modernisation and infrastructure, is bound to wonder whether this is the way forward. "Families displaced from their ancestral village were scattered, forced to rebuild their lives elsewhere. Some sought shelter with relatives, others rented homes, while a few had prepared in advance for this eventuality." The contrast between this and what she says after that brings home the truth of human insensitivity. "As the village sank, people from nearby villages and other areas came to witness the destruction. The submerged homes in Vyasi Lake briefly became the backdrop for selfies".

Dr. Anjan Ray introduced me to **Roli Srivastava**, who had been a student of his. Roli is now a Mumbai-based journalist and founder of *The Migration Story*. She reports on climate change, just transition, gender and migration, focusing mainly on India's marginalised communities. We are grieved when we see ghost villages, the partial ruins of houses and scattered lives, when we see the aged left behind as the young are compelled to migrate towards hope. Her essay begins: "*Kharkhara*, I thought, had a dry ring to it, like the sound of leaves rustling. Or an old man clearing his parched, itchy throat. Or footsteps on the bone-dry farms of this deserted, drought-hit village. Its few inhabitants – the elderly, the sick, the poorest – sat outside their homes, staring at a street lined with padlocked houses".

Most disasters are water-related: storms, floods and landslides; heat waves and wildfires; droughts and waterborne diseases that become more frequent and more intense, during a disaster.

Dr. Roxy Mathew Koll is a climate scientist whose work has advanced the scientific understanding of climate change and weather, influencing disaster management strategies. Some 15 years ago when I first heard him at a climate conference, I noticed his ability to make complex models simple for a layman to understand. I was struck by the passion he brought to climate monitoring and reading weather patterns, and the fact that he could convey it factually, and yet inspire you to want to understand the dynamics. His essay is about the escalating water crisis, monsoon droughts and floods, melting glaciers and coastal vulnerabilities, river interlinking, health implications, and the way forward. He writes, "The Indian Ocean is warming at the fastest rate and the Himalayan glaciers are melting rapidly. As a result, the subcontinent is witnessing a trend of rising floods, landslides, droughts, heatwaves and cyclones. This is affecting the food, water and energy security of the region, deeply endangering lives and livelihoods. India has become the poster child of climate change".

Dr. Somnath Bandyopadhyay is a development ecologist, currently leading a study to develop a River Health Index for the upper Narmada River. In this essay, through narratives which form the warp and weft of wetlands, he brings alive the diversity and uniqueness of particular wetlands. They are the access to fresh water and traditional livelihoods for more than two billion people worldwide, they are ecological hotspots sustaining species from microscopic organisms to large mammals. They help groundwater recharge and erosion control. He writes, "Wetland conservation is about wisdom; not only in how we cherish and restore these living waters, but in how we listen to their silence. Above all, wetland conservation calls for determination, dedication and resolve, in working in harmony with a shared sense of purpose".

Dr. Poornima Jalihal is a senior scientist who heads the Energy and Fresh Water group in the National Institute of Ocean Technology, India. She has coordinated many first-ever projects in ocean energy and desalination with a recognised contribution in the setting up of the first ocean thermal gradient desalination plant at Kavaratti island in Lakshadweep, which helped transform the lives of the island community with its lack of drinking water. She writes, "Mankind will continue to explore deeper depths but any interventions need to be sustainable and keep the ocean safe in its pristine state and grandeur. Sustainable technologies to harness resources in an environmentally friendly manner are the need of the hour. We must commit to protect our oceans while using them in a sustainable manner utilising green technologies".

Mridula Ramesh's books, *The Climate Solution* and *Watershed,* as well as the Sundaram Climate Institute, which she founded, focus on waste and water solutions. She is an active cleantech angel investor. Like all those who have contributed essays to this volume, she wanted the issues related to water to be understood by people outside climate dialogues. She says, "The warmer climate speaks eloquently through water. For India, one of the most vulnerable countries in this climate crisis, adaptation needs to form an important part of the narrative. And in adaptation, water takes centrestage".[1] In her essay, *Philosophy of Water,* she formulates several questions to highlight what our philosophy today is, contrasting it with what it was in the past, when water was respected and managed. "A warming climate accentuates the contrasts of water", she writes, concluding, "Something has got to give. And I don't think it is going to be the warming climate".

When I reached out to **Minni Jain and Philip Franses** after learning about the release of their book, *The Language of Water,* this manuscript was already being formatted. They were both so warm, enthusiastic and committed to this project, they sent their essay at record speed. They focus on how the relationship between humans and the water cycle is the key. "Can we keep the world in balance by learning to speak the language of our most precious resource?"

The Flow Partnership, founded by them, draws from decades of experience with community-led management of floods and droughts in India, Africa, the United States, Europe, the United Kingdom, and other regions of the world, to demonstrate how replenished groundwater can cool the atmosphere, revive local economies, restore food security, store carbon and rebalance our

---

1. https://www.firstpost.com/art-and-culture/in-her-new-book-watershed-climate-expert-mridula-ramesh-stresses-on-urgent-need-of-water-management-in-india-10265231.html

planet. They write, "In times past, the community treated the water with respect and had their own methods of how to hold the water in the land to lessen floods and droughts. That was the norm not the exception. The communities had water embedded into the context of the landscape and their lives. They knew not to cut down the trees, to tend to the soil, to build structures to manage the flow of water if life was to be harmonious. This knowledge of holding water in the landscape was passed on from generation to generation. It is only recently their worldview has begun to be overwritten with one of measurement and control".

The final essay in this collection has been written by me. For the past quarter century, my writing has been primarily about spirituality. Planet Earth is where consciousness has manifested in human form. We are inextricably entwined – the Earth and I. Water is the very basis of life on earth. If we live in unawareness of this, if we forget our connection with all that co-exists here, we are doomed to perish in ignorance. The first step towards conscious living is to be aware of the interconnectedness of everything. If we do not realise this, we would have lost everything, in our own fossilisation. We need to reconnect with the memory of the earth and the memory of water to find the next steps. The way ahead will not come from what has caused the impending systemic collapse. We have to hear the rumble of the Earth, feel the vibrations of water.

This collection of essays is for those who feel the pain of glaciers, rivers, springs, wetlands, oceans. It is an invitation to those who would like to understand why water is the crux of the problem, and how we can heal the earth through water in its different forms. It is a call to all humanity to connect with water – our lifeline – so we can bring life back to a dying Earth.

CHAPTER 1

Rock is Water, Water Rock

Sudeep Sen

IMPRINT

haiku

water's imprint on
  stone, residual — its birth
marks time, age, sex, place.

Sudeep Sen

Sudeep Sen [www.sudeepsen.org] is a leading international poet whose prize-winning books include: *Postmarked India: New & Selected Poems* (HarperCollins), *Aria* (A K Ramanujan Translation Award), *Fractals: New & Selected Poems | Translations 1980-2015* (London Magazine Editions), *EroText* (Penguin), *Kaifi Azmi: Poems | Nazms* (Bloomsbury), *Anthropocene* (Pippa Rann, Rabindranath Tagore Literary Prize), and *Red*. Edited landmark anthologies include: *The HarperCollins Book of English Poetry, Modern English Poetry by Younger Indians* (Sahitya Akademi), and *Converse: Contemporary English Poetry by Indians* (Pippa Rann). *Blue Nude: Ekphrasis & New Poems* (Jorge Zalamea International Poetry Prize), *Rock*, and *The Whispering Anklets* are forthcoming. His photography represented by ArtMbassy, Rome/Berlin [http://www.artmbassy.com/artists.html], is part of private/public collections. The Government of India awarded him the senior fellowship for "outstanding persons in the field of culture/literature." Sen is the first Asian honoured to deliver the Derek Walcott Lecture and read at the Nobel Laureate Festival.

Rock is Water, Water Rock

Sudeep Sen

Rock is water, water rock.
  Stones sing, liquid-splashed.

After heavy rains, granite
  glistens and gleams anew.

Steam stews, slow-simmers
  on its brutal sun-scorched

skin — lines, cracks, fissures
  marking millennia's passing.

Congregation of oval granite —
  gigantic, spectacular, stoic,

unmoved — balanced elegantly
  on impossible fulcrums.

Nature-art's imperceptible
  movements — measured

in tectonic micro-shifts. Like
  blood, aquifers crave life.

Deep embedded cratons —
  slow-carving new gravity.

Erosion

The drops of rain make a hole in the stone,
not by violence, but by oft falling.

— Lucretius

Repetition is an act of great diligence —
the reason stones are shaped the way they are

by wind and water. Erosion's relentless
brush creating rounded, opaque parabolas —

boulders balanced impossibly on pin-points.
Even gravity succumbs — to art-making.

Erasure

Paisleys, florets and oviforms —
  form filigreed lace-like lines

carved by rushes of rain-water.
  Multiple micro-waterways

draw out branch-like tracks
  on red rock-speckled earth.

The streams meet, run parallel,
  seep deep into earth-layers

feeding subterranean wells.
  Over centuries water-patterns

create striations, dips, curvatures.
  Wind-water's erosive power

etch, mineral-mapping granite's
  stone art — erasure's enigma.

Tungabhadra Banks, Hampi

Late evening's deep matt-aquamarine light
slow-merges with the oncoming night.
Mammoth black boulders lie quiet
like sleeping humpback-whales in the shallows.

I can see only one predominant colour-palette now —
an aggregated thesaurus
of lapis lazuli, indigo, turquoise, teal, azure, sapphire,
cobalt, cyan, cerulean —
the colour of the deep blue-black sea,
its wavelength's elongated mystery enveloping all.

From the corner of my eye,
I spot Virupaksha Temple elephant's silhouette
carrying wisdom down Tungabhadra's ghats.
For now, it is relieved to leave behind the elaborate rituals
in the richly carved religious precincts —
its snout partaking pleasures of the river's cool waters.

A local fisherman, secure within the orbit
of his wicker-woven coracle,
paddles across the river, back and forth,
attempting to find the last catch.
It does not matter whether he fishes anything —
just the act of oar-repetition, is yoga enough.

People have long left this place for the day —
the *mandapa* rest-house pavilions empty,
barring few stray dogs.
On the shore, a large family of monkeys
make even the towering peepul tree
groan under their mischievous weight.

At this late hour, I see just remnants —
crumpled flower petals, burnt oil-wicks,
frayed sari-wrapped bundle-offerings,
white-ochre-vermillion marked rocks worshipped as gods.
River gods come in various *avatars* —
a snarling yellow-striped tiger, an obsidian Nandi bull,
small statuettes of marble Ganesh,
an array of carved Shiva *lingams* on wet granite
embedded in a mosaic of *yonis,*
and slews of oval stones.

Imminent night's low light deceptively hides
what you do not wish to see —
pilgrim's litter, their desecration, their misplaced prayers.
Despite graffiti's defacement,
I can read Vijayanagara's grand stone-etched histories,
hear the whispered medieval chants
decipher the inscriptions' coded clarity —
solitude, serenity,
and the simplicity of somnolent seance.

Shiva Lingam

Protected in a rock-chambered
temple, a gargantuan granite

shiva lingam stands, floating
on water. It is a miracle

of buoyancy — an impossibility
made possible by a mirage.

Early-mornings, late-evenings —
I see an old bare-bodied priest,

a sacred Brahmin-thread strung
across his frail bony-chest,

clad in knee-length frayed *dhoti.*
He climbs down towards

the *lingam*'s sunken plinth.
Wading to the waterline base,

he ascends the monolithic tiers
where the female *yoni* meets

the male *lingam*. On it, he puts
fresh marigold petals, milk —

applies sandalwood, turmeric
and vermillion oil-paste.

It is a ritual — not of marriage —
but a union of the elements.

But how does the waters
of the Tungabhadra River —

many kilometres away —
reach its base everyday

with a fresh flood of aqua?
Much remains unexplained —

subterranean aquifers, natural
waterways, channel-tunnels,

stepwells — how they were
hand-hewed over generations

by our ancestors. Enigma and
faith, go hand-in-hand —

I am happy with this mystery.
I offer a simple prayer,

toss a talismanic coin and raw
crushed-earth into the pool.

As they make circular rings,
its sonar-echo carries back

in its sine-wave rhythms —
a song in an ancient tongue.

If ever I learn this forgotten
  language, I might be able

to decipher its script, its story.
  Life's music is sung by all —

only its deep-coded *mantra*
  has the power to move rocks.

Relentless Rain Beats Down Through The Night And Day

Relentless rain beats down through the night and day —

it is unceasing — balletic in its falling. Silver striations

of wet-streams cast impressionistic paintings

on glass window-panes. Endless thunder-showers —

a choral refrain. Rain's *fullness* — a fecund presence

Lake Haiku 1

water's mirrored sheen —
swan's skate, gliding on its skin —
marine undertow

Lake Haiku 2

clouds lie on water
photo-perfect and stone still —
a transient frame

Red Rain

Chopped wood —
  stacked, floating dead.

An idle boat —
  rudderless, adrift.

Black rain,
  red rain —

but not a drop
  to drink or taste.

No space left
  for humans —

only beauty
  of contamination

like rainbows
  and sunsets —

prolongs illusion.
  It is illusion

that allows us
  to live, dream —

charting out our own
  perpetual diary.

Night Water

Galactic plasma, cosmos's green blood —
night liquid lit by a light mimicking a distant sun.

Pre-natal amniotic fluid, embryo swimming
in a womb, pre-birth — before emerging whole.

Crickets count our heartbeats, as a lone-bird tweets
a mother's call. You walk on water, chest high,

your garment soaked and translucent. You pirouette
subaqueous, limbs half-swimming, fingers fluttering

like a wingless moth seeking glare's heat in this cool,
hands circling a patterned dance-form — your eyes

peering into darkness excavating the gathering stars,
their optics hidden to human eyesight's secrets.

This may sound like silence — a Sapphic fragment,
a delicate phrase of an *alaap* or overture packing

multiple melodic strains in one sequence, a half-pause.
As pure sonics of the rumbling Milky Way reveals

itself for a millisecond — you offer your obeisance
on the water's edge, as its surface-tension melts

and eases for a moment, allowing for the intricate
abhinaya's gliding arc to close her parenthesis in.

i.e. [That Is]

i.e.

because you hear —
the sound
of a lone rustling leaf —
you hear the sea.

i.e.

because I consider
the sea silent —
you hear its silence
in my studio.

i.e.

& because of that —
the silence will not empty
the sea,
of its leaves.

Disembodied

1.

My body carved from abandoned bricks of a ruined temple,
from minaret-shards of an old mosque,
from slate-remnants of a medieval church apse,
from soil tilled by my ancestors.

My bones don't fit together correctly as they should —
the searing ultra-violet light from Aurora Borealis
patches and etch-corrects my orientation —
magnetic pulses prove potent.

My flesh sculpted from fruits of the tropics,
blood from coconut water,
skin coloured by brown bark of Indian teak.

My lungs fuelled by Delhi's insidious toxic air
echo asthmatic sounds, a new vinyl dub-remix.
Our universe — where radiation germinates from human follies,
where contamination persists from mistrust,
where pleasures of sex are merely a sport —
where everything is ambition,
everything is desire everything is nothing.
Nothing and everything.

2.

White light everywhere,
but no one can recognize its hue,
no one knows that there is colour in it — all possible colours.

Body worshipped, not for its blessing,
but its contour —
artificial shape shaped by Nautilus.

Skin moistened by L'Oreal
and not by season's first rains —
skeleton's strength not shaped by earthquakes
or slow-moulded by fearless forest-fires.

Ice-caps are rapidly melting — too fast to arrest glacial slide.
In the near future — there will be no water left
or too much water that is undrinkable,
excess water that will drown us all.
Disembodied floats, afloat like Noah's Ark —

no GPS, no pole-star navigation, no fossil fuel to burn away —
just maps with empty grids and names of places that might exist.

Already, there is too much traffic on the road —
unpeopled hollow metal-shells without brakes,
swerve about directionless — looking for an elusive compass

Burning Ghats, Varanasi

Over-heated flaming pyres of the burning dead,
partially shields my sight of river Ganges —
its fast muddy currents eddying the floating lamps, bathing bodies,
remains of corpses, flesh-bone ash.

At Manikarnika Ghat, a mixture of sanctity and stench
rises from silted sands and wooden armatures —
fire-aided decomposition of human flesh —
the offerings swiftly lapped up by roaming animals.
An emaciated *sadhu* with wild-knotted dreadlocks,
perched precariously on a bamboo frame
on the edge of the river,
dreams of alms that might come his way, even at this late hour.

Presiding priests, feed ritual *ghee*
to the burning wood-and-dead —
its flames forming huge flares,
fragmented waves of golden-amber spark,
electrifying helical fire-flurries — a living, crematorium drama.

A young boy scratches his newly-shaven head,
a pot-bellied man immerses himself in the river,
stray dogs bark, cows groan, loudspeakers bray.
Gandhi's posters on ghat-side walls preach peace, non-violence.
Amid so much noise,
the business of death being transacted
carries on, without any emotion or fuss.

Saffron-robed men on *ghat*-steps
sit in yoga postures, praying —
a silent quest —
what does prayer amid all this din and commerce get you anyway?

Medley of bells, conch, chant, fire, water, boat, people
ceases to be a cacophony after a few hours —
variant decibels melding into a drone, a trance —
where the only balance that exists, is in our minds.

Bare-headed, bare-bodied young men, draped in swathes of pure cotton,
foreheads smeared in sandalwood and vermillion
carrying ash-filled earthen pots —
walk past me towards the river-edge,
detached — eldest sons performing last rites for their dead.

White-clad teachers squatting cross-legged on the ghat
under large circular cane-parasols
impart teachings from the *Upanishads* and *Vedas*
to young priests-in-the-making.

Illuminated cane-lanterns
hang on long bamboo poles curving skywards —
homage to the memory of martyrs —
guiding light for heavenly *apsaras*
descending during the Kartik month to bathe in Kashi —
as oil-soaked wicks flickering on beds of rose petal, sail
catching the waves' moods.

In the super-heated pyre, I hear another ritual pot break,
another skull crack, another soul take flight.
I see some shore-temples slow-sink into the swallowing river —
effects of unpredictable tides and climate change
taking with them, both the mortal and the immortal —
ashes to ashes, dust to dust —
water to heavy water, life to after-life.

Rising Sea Levels

The ever-hooded, tragic-gestured sea.
— Wallace Stevens

Granite outcrop
  that once jutted out
of the ebullient sea —

fifty metres
  from the shore —
is seen no more.

The lighthouse —
  our beacon,
an adventure island.

We'd swim to its base —
  shells, mussels,
striated oval stones,

seagull eggs nested
  on slippery
kelp-laced rocks.

As children, reality —
  as adults,
a submerged memory.

Drought, Cloud

It is bone-dry — I pray for any moisture
  that might fall from the emaciated skies —

There is a cloud, just a solitary cloud
  wafting perilously —

But it is too far in the distance
  for any real hope — for rain.

CHAPTER 2

India's Aquifer Story

Himanshu Kulkarni

"If there is magic on this planet, it is contained in water."
~ Loren Eiseley

Himanshu Kulkarni

Dr. Himanshu Kulkarni is a Founder Trustee, former Executive Director and currently Scientist (Emeritus) with ACWADAM, a not-for-profit institution working on groundwater since 1998. He led ACWADAM since its founding, until 2022, as Executive Director. Dr. Kulkarni is also associated with three leading academic institutions, namely as Professor of Practice for the Rural Management Programme at Shiv Nadar Institute of Eminence in Greater Noida, as Visiting Professor at Ashank Desai Centre for Policy Studies. He also teaches for the Water Policy and Governance Programme at the Tata Institute of Social Sciences in Mumbai.

Dr. Kulkarni has been working on aquifers and community action across India's diverse groundwater typology for more than forty years. He has held advisory positions on various committees of the Government, apart from anchoring many action-research collaborations on groundwater, including international programmes. Dr. Kulkarni has published his work, while also working with communities at the grassroots, demystifying groundwater science and catalysing aquifer-based participatory groundwater management and governance. He has supervised, guided and mentored students from many institutions and disciplines on the singular subject of groundwater for many years in various positions. Dr. Kulkarni continues to work on bringing a strong 'aquifer' based pedagogy into various dimensions of water management and governance in India.

India's Aquifer Story

Himanshu Kulkarni

Introduction

The global annual extraction of groundwater is of the order of nearly 1000 km^3.[1] This number, incidentally is also comparable to the total groundwater depletion over the past century – of about 1000 km^3 per year.[2] By the turn of the century, South Asia alone had more area under 'groundwater irrigation' than all the rest of the world combined, with the early period of the last decade indicating that 30% of the global area under irrigation was in South Asia.[3] India now is the largest user of groundwater resources in the world, as indicated by many different studies[4], even though global data sets indicate at least eight other nations to have a higher per capita annual groundwater consumption. In fact, by the mid-1980s, India had become the largest groundwater user in the world, which may have prompted the process of periodic groundwater assessments by the Central Ground Water Board (CGWB) – the apex agency dealing with groundwater under the Ministry of Jal Shakti of the Government of India – from 1987 onward.

1. Margat and van der Gun, 2013
2. Scanlon et al., 2023
3. Shah, 2009
4. Shah, 2009, Scanlon et al., 2023

Given India's great dependency on groundwater in agriculture, for rural domestic water supply and almost equal share with surface water in urban water supply, many regions have witnessed groundwater extraction in excess of the potential annual recharge to aquifers, leading to a situation of exploitation, although the commonly used term to express the degree of exploitation is 'overexploitation'. In many regions of the country, groundwater contamination has also emerged co-terminus to acute groundwater scarcity caused due to groundwater overexploitation. Aquifer management holds the key to resolving India's current groundwater crisis because of the sudden shift to overextraction of groundwater (seven decades) from a long legacy of systematic management through sustainable approaches to extraction (several thousands of years).

The Current State of Our Aquifers

The latest national assessment of the groundwater resources of India shows that 26 to 32% of the units (blocks and talukas) are in the 'unsafe' category, where extraction to recharge ratios are above 70%.[1] While the percentage seems small, local scales of assessment might tell a different story and there are villages and watersheds even in safe talukas that indicate long-term depletion or short-term severe scarcity of groundwater resources, even in seemingly 'safe' categories of the groundwater assessment in locations such as Central and Eastern India (Patil et al. 2021).

The latest report on groundwater quality produced by CGWB (2024a) indicates that groundwater contaminants are not just prevalent in areas with serious groundwater exploitation but are also emerging in areas where the groundwater assessment units fall in the safe category. A quick district level synthesis of the

1. CGWB, 2024, 2022

two latest CGWB reports (CGWB, 2024a and 2024b) shows that out of 712 districts, only 120 districts are 'safe' and do not have any of the seven measured groundwater quality contaminants which includes EC, Chloride, Fluoride, Nitrate, Iron, Arsenic and Uranium. Thus, the remaining 83% of the districts fall under the 'not-safe' category, from a groundwater overextraction limit, or have seen the occurrence of at least one of the seven parameters profiled for groundwater quality assessment.

India's current crisis of groundwater has three obvious consequences:

1. Depletion of groundwater indicated by long-term groundwater decline in wells.
2. Contamination of sources – wells and springs – especially sources of drinking water.
3. River flow depletion brought about by the depletion in base flows of streams and rivers as a consequence of decrease in natural discharge of groundwater to different water bodies.

The above consequences lead to what can be called the 'impact of impacts'. Economic impact such as drop in incomes from agriculture, further marginalisation of the marginalised especially regarding safe drinking water access, slip-backs in public water supply systems and a growing gap between increasing demand and public supplies of water are examples of such impacts (Kulkarni et al., 2014). What is emerging as one of these profound impacts is the effect of groundwater contamination on human well-being and health.

While agriculture accounts for much of the total usage of water in India, urban water footprints (water used over unit areas) are often larger than those in agriculture.[1] Nearly

1. Kulkarni et al, 2023

half of urban water usage is derived from groundwater.[1] Most discussions on urban groundwater focus on the issue of reduction in infiltration-facilitation surfaces and increased extraction. As cities grow rapidly, the tendency to build vertically, implies deeper foundations including basement amenity spaces that often require digging out the (shallow) aquifer material that was originally storing and providing groundwater supplies on a regular basis. In Pune city alone, at least a million m^3 of shallow groundwater storage has been permanently lost as such basements and foundations replaced parts of the shallow basalt aquifer[2]. Further, while precise estimates are not available, large volumes of groundwater have been pumped out – often into natural drains and gutters - from basements and foundations both during construction and after completion of buildings, simply a wastage of a valuable resource that can be put to use amidst the growing crisis of scarcity of groundwater resources in other parts.

Most discussions on groundwater are largely centred around the large-scale exploitation of groundwater resources, rather than being aquifer-centric. Watersheds and river basins have been in discussions as socio-hydrological units in managing water from small areas (microwatersheds) to larger catchments at river-basin scales.[3] On the other hand, mapping of aquifers as systems of managing groundwater resources have found importance in India only during the last couple of decades[4] with CGWB's NAQUIM Programme – as summarised in CGWB(2024), being one of its kind in the world.

Given the current crisis of groundwater and the need to recognise and manage groundwater as common pool resources,

1. NIUA, 2011; Narain, 2012
2. Kulkarni et al., 2023
3. Shah, 2013; Tare, 2019
4. Kulkarni and Vijayshankar, 2009

'aquifers' must become the focus of governing groundwater resources, in planning processes of recharge, conservation and in managing extraction. Aquifer understanding will also help address questions regarding the variability and changes in interconnections between groundwater and various ecosystems that constitute India's diverse geography in a changing climate.

India's Groundwater Legacy: Natural Springs and The Large Diameter Dugwell

With a few exceptions, groundwater extraction around the world remained from relatively shallow depths using human and animal traction – mostly through the diversion of spring-water, dugwells or by systems such as the qanats – until the end of the 19th century, or even much later in some countries; knowledge on the presence of any aquifers beyond the near surface was in most cases non-existent.[1] The oldest dugwells in India date back thousands of years, e.g. the dugwell discovered at the archaeological site at Dholavira in Kutch. Uses of wells driven by human or animal power goes back many centuries in India. So does the access to natural springs as experienced by the author during his study travails across the Himalayan landscape. Traditional systems of groundwater access, extraction and distribution are reflected through the *rahat, mhot, chadas, dheklas* and *tendlas*. Our own version of the *qanat,* called the *suranga,* is still in use in some parts of the country.

A remarkable aspect of this centuries-old pattern of groundwater usage is its diverse application across various geographies. From arid desert regions to high-altitude mountainous areas, groundwater has served as a lifeline for both humans and ecosystems. In many of these regions in India,

1. Moench et al., 2012

groundwater has played a vital role in supporting life and enabling livelihoods. The ubiquitous occurrence of the shallow dugwell and the appropriate mechanisms involving animal and human traction for such extraction, bear testimony to a well thought out knowledge system of sourcing and strategic design of extraction. The *rahat,* for example, is based on the design stemming from the Persian Wheel that was used in large parts of Asia and Africa for many centuries; it works on the principle of steady-state water levels in a dugwell[1], and can be adapted to conditions in the aquifer that the well taps, from season-to-season and across diverse settings of aquifers.

Thus, these traditional mechanisms of shallow dugwells, fitted with the 'Persian Wheel type' of extraction systems, must have needed an understanding of at least the local conditions in the shallow aquifer. Similarly, access to springs through a variety of locally suitable systems has been recorded, especially in the mountains, for many centuries. The use of groundwater emerging from *chashmas, chhumiks, mools, dharas, naulas* and *baoris* – as they are referred to locally in different regions – also bears testimony to the different ways in tapping spring-water, depending upon the location, the nature of usage and the patterns of spring discharges, the latter entirely depending upon the behaviour of groundwater in the underlying mountain aquifers, especially upon the discharges of springs, which are a reflection of the storage and transmission characters of the underlying aquifers.

Hence, India has been as much a groundwater civilisation as it has been a river valley civilisation! Uses of wells driven by human or animal power goes back many centuries. So does the access to natural springs in various forms. Conservation and infiltration facilitation mechanisms listed under CSE's

1. Kulkarni et al., 2018

traditional rainwater harvesting examples bear testimony to the customisation of these long-standing examples to climate, hydrology and sub-surface conditions.[1] While traditional means and mechanisms of groundwater usage continue in some parts of the country, many of them have disappeared or seem to persist only as relics of this long-standing legacy, especially in dealing with the shallow, sustainable systems of aquifers in different parts of the country. The transitions away from these age-old systems of groundwater usage practices is not just about the modernisation of groundwater extraction but represents a huge socio-ecological shift around perceptions and practices of groundwater sourcing and access, especially during the last fifty years or so. The transition from community practices to 'individualisation' of the resource has emerged slowly since the 19th century and has accelerated during the last four to five decades. The growing numbers of wells and increasing depths of wells are clear evidence of this shift. Deep boreholes and tubewells are now attaining depths in close to a thousand metres, as against depths that were in single digit metres about half a century ago. The most significant change, though, is the abandonment of fundamental concepts that balance the hydrology of aquifers and the choice of appropriate technologies that lead to sustainable practices such as the steady-state functioning of well-aquifer management in the Persian Wheel concept. A rapid shift to a solely technology driven definition of extracting water – irrespective of the consequences – over short periods of time, is one of the reasons for the rapid decline of groundwater levels and the iniquitous impacts on wells even in a single village.[2]

---

1. CSE – www.rainwaterharvesting.org
2. Kulkarni and Vijayshankar, 2014; Kulkarni and Gokhale, 2021

The Longer History of Groundwater

The story of the dugwell in India, represents a centuries-old story of how humans interacted with aquifers, adapting to local conditions of geology, climate and other socio-economic drivers prevalent at different periods during a long history of thousands of years, coming into the present. This history reflects the broader canvass on how humans engaged with water in general and groundwater in particular through different periods of the recent geological past in which humans evolved not only as living beings but also in terms of how they perceived, sourced and accessed water. Burke at al. (2012) provide a narrative of this record that has been summarised below:

1. 2 million years until 14000 years BP: hunting-gathering for food and water, often living close to perennial supplies from springs, shallow ditches and tapping base flows in river-beds.
2. 14000 – 13000 years BP: transition from foraging to sedentary agriculture – beginning with systematic localised water management for growing food and domesticating animals.
3. 11000 – 9000 years BP: settlement patterns determined by water availability – techniques of excavating wells probably linked to discovery of flint and metallic minerals.
4. 2500 – 200 BP: Technological rapid advancements for water access and extraction based on basic principles of physics and engineering, e.g. the concept of levers in developing extraction mechanisms using poles / long pieces of wood / bamboo such as those observed in the case of the *lattha* or *tenda* in Eastern India; or the

systems based on the concept of the Persian Wheel – the *rahat* or *mhot* or *chadas* – that use the technology of gears and pulleys for extracting water, even if it is through human and animal power. Some of these systems are still in operation and have survived pressures from modernisation even today.

5. 19th century onward: Rapid development in water resources engineering – whether in the form of large water infrastructure (dams and canals) or technologies of drilling deep into the earth for access to deeper groundwater resources.
6. Last 50 years: Booming water extraction with innovations in energy – driven by rapid growth in the demand for water – variable technologies that allow multiple points of extraction at great rates and the capacity to convey water, both, over long distances and (lift it) to great heads (heights).

Parts of the present are often a peep in the past realities of the past. The Adivasis in India, have been living in the hilly, forested and often in some of the most inaccessible niches of India's diverse landscape and still follow some of their traditional systems of using and managing water, rather frugally. Springs and shallow wells are the mainstays of community water supplies, often through innovative systems of distribution that are prevalent in large regions of Adivasi dominated geographies such as on the Eastern India Plateau and the Northeastern Hilly Regions.

Bhave and Kulkarni (2023) highlight the above aspect stating the following: "As talks of environmental issues like climate change gain momentum, tribal communities that dwell closest to Mother Nature have become the most vulnerable to natural forces on one side and a plethora of anthropogenic fluxes related to the accelerated growth engine of the modern world on the other. As an example, spring water clearly holds a place in the lives and lifestyles of many ethnic communities in India. Many of these communities have a history of thousands of years of living in these regions and carry a legacy of sustainable and equitable water management that not only needs to be emulated but needs protection from the externalities of modern day intrusion."

Shallow dugwells, despite showing a decline in proportion to the growth of tubewells and boreholes,[1] still persist and are still significant in numbers in districts in the state of Maharashtra that is underlain by hard-rock aquifer systems from the Deccan Volcanic Province, perhaps the seemingly least likely location for dug wells. However, they are also prevalent in districts from Tamil Nadu, Gujarat, Rajasthan, Telangana and Andhra Pradesh. Most dugwells are individually owned but are still group-owned and used. Hence, the shallow aquifer systems, at least in the hard-rock regions of India still provide recourse to farming, not to mention their usage in urban centres for which there are few reliable estimates. In many ways, this pattern represents the continued faith and trust in the legacy of several thousands of years upon the shallow aquifer in the Indian subcontinent.

The History of Aquifers

Groundwater is a *fugitive* (cannot be held 'captive' underneath a defined piece of land) and *invisible* (as a subsoil resource it is largely unseen) common pool resource.[2] We can safely surmise that the CPR or 'commons' nature of groundwater has steadily changed to a case of individual access leading to the tragedy of

1. MI Census, 2017
2. Blomquist and Ostrom, 1985; Ostrom, 1990

the commons described by Hardin (1968). The paradox of increasing dependency and deepening vulnerability is emerging across the villages and towns in different parts of India; and so are the responses to such problems. A brief list of the various programmes and initiatives through which groundwater problems are being addressed is provided as a table below:

Main activities focusing on groundwater management in India	Specific linkages to groundwater management and governance	Organisations involved	Larger government schemes involved
Augmenting groundwater stocks through watershed management	Largely in the regions underlain by undulating topographies and underlain by consolidated rock formation	Civil Society Organisations (CSOs) State Government Departments like Rural Development, Agriculture etc., Corporate Social Responsibility (CSR) initiatives	PMKSY, MGNREGA
Augmenting groundwater stocks through injection of rooftop rainwater or diverting runoff from land surfaces and surface water courses	Urban areas and in flatter landscapes such as the alluvial regions of the flood plains	State Departments, Urban Local Bodies, Private Initiatives, Industry and CSOs	Shallow Aquifer Management Pilots of MoHUA; Artificial Recharge schemes of CGWB and even ABHY

Springshed management for the revival of springs in the mountain regions	Mainly the Himalayan Region but also in parts of the Western and Eastern Ghats and other highlands	Largely led by CSOs in partnership with State Governments; some initiatives by international donors like ICIMOD and SDC	Many programmes in progress, almost across all Himalayan States – Meghalaya, Nagaland, Sikkim, for instance – PMKSY, MGNREGA
Managing demand for water in agriculture through shifts in cropping	Limited locations but at scales of villages, watersheds and smaller locations largely in rural areas through programmes like participatory groundwater management (PGWM) or through larger initiatives of Government Programmes	Mainly Civil Society Initiatives and State Initiatives in agriculture	PMKSY, National Millets Mission, Natural Farming, Atal Bhujal Yojana
Regulatory functions through formal legislation such as limiting the numbers and depths of wells	Some states are implementing legislations on groundwater, mainly through regulation of the depth of boreholes and tubewells	The State	State Government Acts and Legislation or through special notifications by district administrators

Using energy to regulate groundwater	Separation of feeders for irrigation and rural domestic demands but in limited areas, although alternatives such as solar pumps are being	State Electricity Regulatory Institutions and other such bodies	State Energy Regulatory Departments
	introduced to check groundwater extraction		
Source sustainability of drinking water sources	Protection of sources from overexploitation in other sectors; capping pumping of irrigation sources during periods of stress	State and district administration	JJM through the Ministry of Jal Shakti

Table: Summary of issues surrounding groundwater that are being addressed through different organisations and programmes in India

A crucial point for reflection from all the initiatives summarised above and the foregoing discussion is the common reference to groundwater in the context of:

1. Extraction
2. Quality
3. Recharge
4. Management
5. Governance

What is largely missing in the mainstream discourse on groundwater is the reference to aquifers. While surface water discussions are commonly referred to with reference to watersheds and river basins, any reference to groundwater and groundwater management must find a stronger reference to aquifers, the resource systems for groundwater units in the parlance of a common pool resource. An aquifer (a sub-surface rock formation that can store and transmitting groundwater) is a hydrogeological unit that helps us understand how groundwater resources behave in space and time. Aquifer systems allow for better understanding on the interconnections between different parts of the water cycle with aquifers, including the connection between overlying watersheds and underlying aquifers. Springshed management, for instance, is based upon understanding this relationship and is the foundation for improved systems of managed aquifer recharge in mountainous regions.[1]

While we often talk about the age of groundwater and the history of groundwater usage over the last 15000 odd years, it is time to embed the history of aquifers into future discussions on groundwater. At ACWADAM, the organisation that I represent, we have tried through the last twenty years, to develop a broad typology of aquifer settings in India, that is classified into six different types. While this has been extensively used in many previous articulations, we present it here to complete India's aquifer story (Figure 2, Table 2).

1. Shrestha et al, 2018; Kulkarni et al., 2021, Tambe et al., 2020

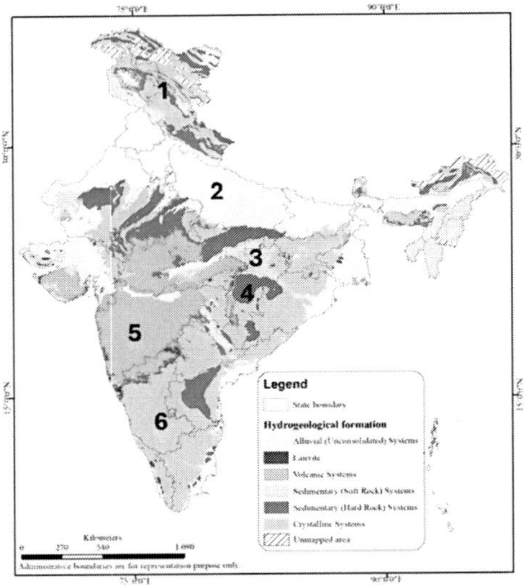

1	The Himalayan melange (mixed type of rocks) and associated aquifers	Unrecognised as compared to aquifers in the plains, these complex systems of localised aquifers provide water to habitations and to ecosystems, mainly through spring discharges	2000 Ma to recent
2	The unconsolidated (largely alluvial) aquifers	Regional aquifers lying largely in the flood plains and delta regions, representing one of the most used groundwater systems in the world – regional aquifers each often supporting many habitations through tubewells	A couple of Ma to present
3 & 4	The consolidated (soft and hard) sedimentary aquifers distributed across many regions	A mixed set of aquifer systems spread over different parts of India, concentrated across a longitudinal spread in west, central and eastern parts of India and are often in conjunction with forests and mining areas	1700 Ma to tens of thousands years

5	The basalt aquifers of the Deccan Volcanic Province	Located in west-central India, these are the most	About 65 Ma
		heterogeneous aquifers in the country, tapped largely by dugwells and supporting both rural and urban Indian water supplies; also hosting spring systems from the northern parts of the Western Ghats	
6	The crystalline (rocks of igneous and metamorphic origin) aquifers dominantly in peninsular central and eastern India	Aquifers that are spread across large parts of eastern and southern regions, quite inhomogeneous in nature and often producing water from shallow depths; deeper drilling in some regions has also revealed water stocks at depth	4500 to 500 Ma

The Way Forward

Aquifer depletion, groundwater contamination and the tension created between aquifers and ecosystems, especially rivers, often go together. The shrinking size of landholdings; the decentralised nature of governance required to address the largely atomistic nature of groundwater extraction and its impact; and the diverse socio-hydrological aquifer settings makes it difficult to implement top-down approaches to groundwater management, including the often-obvious instrument of groundwater legislation.

Local, participatory aquifer management and governance involves a paradigm shift on many fronts. This paradigm attempts to move the focus from sourcing of groundwater to that of resource (aquifer) management using the principles of managing groundwater as a common pool resource through a protocol that is based on social hydrogeology, a concept that is gaining importance across the globe. The commonality, across many local efforts, lies in the strength of community members participating in the process of developing local knowledge, arriving at decisions, and building actions around such community-based decisions. In most cases, they include mechanisms of decentralised groundwater governance as part of the larger groundwater management strategy. At the core of all these attempts lies a combination of 'a transdisciplinary approach and the micromanagement of groundwater problems', a combination that holds promise in the pursuit of not just the SDGs, wherein the aspect of groundwater remains implicit but crucial to their achievement, but also in otherwise unseen dimensions such as building disaster resilience.

Such efforts also hold promise for an improved dialogue between groundwater science, praxis, and policy. Participatory aquifer management requires a strategic combination of demystified science, participatory practices, and robust policies, all of which embrace the process of ensuring a seamless integration of efficient and equitable groundwater management with sustainable groundwater governance. The process must, to begin with, include demystified science leading to the development of knowledge, data, skills and understanding of groundwater resources at any given location. Based on such understanding, community-level decision-making, co-operation between stakeholders must be supported through robust systems of governance that include socially normative regulation backed by formal legislation, which together will define the institutional structure of groundwater governance in India. Participatory processes of management and governance of groundwater can lead to a socially normative protocol where community decisions are documented and accepted through institutional instruments such as resolutions of the village gram-sabhas (local instruments of governance under the gram panchayats—village levels elected democratic institutions of governance vested under the Constitution of India).

Combining social norms and formal legislation is challenging but holds the key to bring policies closer to the ground and enable sustainable practices to outstay many different externalities. The concept of decentralised governance, given that the actual decision-making on groundwater will be vested at the village or sub-district levels, becomes quite relevant to India's atomistic groundwater problem. Reconciling legislation to the nuances of a decentralised normative framework will be challenging, but keeping the long-history of aquifers at the centre of such legislation could be a novel way of not just managing and governing groundwater resources but also ensuring that aquifers, like other ecosystems, remain protected in ensuring sustainable groundwater resources to meet the water needs of the future.

CHAPTER 3

Reviving Life's Source

Sanjeev Bhuchar

"Eventually we will realise that if we destroy the ecosystem, we destroy ourselves."

~Jonas Salk

Sanjeev Bhuchar

Sanjeev Bhuchar serves as the Senior Intervention Manager – Springs, at the International Centre for Integrated Mountain Development (ICIMOD), where he plays a pivotal role in efforts to revive and manage springs sustainably across the Hindu Kush Himalaya (HKH) region. ICIMOD's springshed initiatives have not only advanced scientific understanding but also empowered local institutions and stakeholders – particularly women and marginalised groups – to play active roles in the planning and stewardship of water resources.

With nearly three decades of experience in the fields of water resources and watershed management, he brings technical expertise, regional knowledge and a strong commitment to mountain ecosystem restoration. His work contributes to developing and scaling science-based springshed management approaches that are community-centric, inclusive and resilient to climate change. Sanjeev has been contributing to building regional partnerships, capacity development programs and policy dialogues that prioritise the protection and conservation of springs as critical lifelines for mountain communities.

Reviving Life's Source

Springshed Management in the Hindu Kush Himalaya

Sanjeev Bhuchar

Silent Lifeline of the Mountains

Imagine waking up in a small Himalayan village with the sun rising over the towering peaks, painting the landscape in golden hues. For generations, communities here have relied on crystal-clear water from natural springs, called by many names such as *dhara, naula, bowdi, paniyar and chashma*. This water has quenched their thirst, helped their crops grow and supported their daily lives. Springs scattered across the Hindu Kush Himalayan mountains spanning Afghanistan, Bangladesh, Bhutan, China, India, Nepal and Myanmar are not just water sources. They are lifelines, woven into the region's culture, economy and environment.

But this lifeline is in trouble.

Precious Resource

The Hindu Kush Himalayan region, home to about 240 million people, boasts nearly 9 million springs as an approximate estimate. These springs supply water for drinking, washing, bathing, cooking, farming, biodiversity and ecosystems, making them critical for survival. In Nepal, for instance, over 10 million people depend on them, while in India, about 50 million residents in the

Himalayan region rely on the estimated 3 million springs scattered across its terrain, according to a report published by NITI Aayog. Even bustling urban centres like Mussoorie in India and Murree in Pakistan, depend heavily on these natural fountains.

Springs are not just sources of water. They are part of the region's cultural identity. In Bhutan, India and Nepal, many springs are considered sacred and linked to spiritual legends. They are also havens for biodiversity and crucial for feeding rivers, wetlands and wildlife. But beyond their cultural and ecological importance, they play a very practical role: they keep life flowing in the mountains.

Vanishing Springs

Over time, many springs have begun to dry. In India's Himalayan region, almost half the perennial springs have become seasonal or disappeared altogether. Similar trends are seen in the Chattogram hill tracts of Bangladesh, across Bhutan, Nepal and in the mountainous regions of Pakistan, where the drying up of springs forces communities to walk farther for water and sometimes even consider moving away. In some places, like Bhutan's Lholing village in Paro district, the disappearance of springs has driven people to migrate in search of water.

The reasons are complex. Climate change, with erratic rainfall and longer dry spells, plays a big role. Poor land use practices, deforestation, unplanned construction, loss of traditional practices like the buffalo-wallowing ponds which also serve as recharge ponds, deep boring and overuse of groundwater as well as earthquakes add to the problem. The demand for spring water is rising steadily as residential population, as well as tourism in the HKH region continues to grow, driven by the influx of visitors seeking the region's natural beauty, adventure activities and the cool climate during summers. Imagine a glass slowly draining while more and more people try to sip from it - that is the situation for many mountain springs today. It is not just villages that rely on springs for their water needs. Darjeeling, with a population of over 118,000 (Census of India, 2011), receives its entire municipal water supply from 26 springs. Similarly, in the Indian Western Himalaya, Mussoorie's water supply depends on 20 springs. The drying of springs poses a significant threat to the future of these hill towns, especially during the tourist season when demand is high, but water supplies are low.

The quality of spring or spring-fed stream water is deteriorating due to increasing environmental pressures. Improper disposal of solid waste and untreated sewage has led to contamination of natural water resources, while the growing use of chemical fertilisers in nearby agricultural activities contributes to nutrient pollution. These pollutants seep into groundwater and springs, disrupting the natural balance and reducing water quality. The situation is further exacerbated by the lack of proper waste management systems and awareness about sustainable practices, posing a significant risk to local communities and the diverse ecosystems dependent on these water sources.

Women and Children Bear the Burden

In many mountain communities, fetching water is a task traditionally assigned to women and children, reflecting deeply ingrained social and cultural norms. This task, already labour intensive and time consuming, becomes increasingly challenging as springs dry. In regions like Nepal, studies show that nearly 80% of people fetching water are women, underscoring the burden of water scarcity.

As springs deplete, women and children are compelled to walk longer distances over rugged terrain, often carrying heavy loads of water. This physical strain not only exhausts them, but is also a health risk, causing musculoskeletal injuries and chronic fatigue. For children, particularly girls, these extended journeys frequently disrupt their education. Time spent on water collection replaces what could be used for studying, attending school, participating in extracurricular activities.

For women, the increased burden limits opportunities to engage in income-generating activities, or even personal rest and leisure. This cycle perpetuates gender inequalities, as the time and energy spent on water collection reinforces their traditional roles and their ability to break out of poverty. Addressing these challenges requires investing in local water management systems, rejuvenating natural springs and fostering resource distribution to alleviate the burden of women and children in mountain communities.

What Can Be Done?

The good news is that solutions exist. Scientists and local communities are working together to better understand how springs work and how they can be revived. Springs are not fed by surface water alone; they are connected to underground water systems called aquifers. Protecting these aquifers, through measures like reforestation, restoring traditional ponds and improving land management, can help rejuvenate springs.

In some areas, like Sikkim in India, traditional practices are being revived to recharge groundwater. Local projects in Nepal and Bhutan are testing ways to conserve springs and improve water quality. In Mussoorie, for instance, springs still provide nearly 9 million litres of water daily, thanks to careful management.

The six-step protocol for reviving springs published by the International Centre for Integrated Mountain Development (ICIMOD), provides a practical and systematic approach to revive springs. Here is a simple breakdown of the protocol:

1. Identify and map springs: Locate and map the springs in the area, understanding their significance for the community and environment by engaging local communities.
2. Study the spring's characteristics: Observe how the spring functions, its flow patterns, as well as the seasonal variations in flow and water quality to determine its current condition.
3. Understand the gender, social and governance systems: Recognise how water access affects different groups, particularly women and marginalised. Note if there are any formal or informal management of the springs and how inclusive these institutions are.
4. Identify the recharge area: Identify the area where a spring gets its water (recharge zone) by analysing the geology, soil and landscape.
5. Plan and implement interventions: Co-design solutions to protect and restore the recharge zone. Plant trees, preferably native species that aid in holding water in the soil; reduce erosion; gully stabilisation and building structures to increase water infiltration.
6. Monitor and Maintain: Regularly measure the spring's flow and condition, making required adjustments to maintain its health over time.

In the quiet village of Moldhar, Uttarakhand, nestled in the Indian Himalayas, a group of women are leading a transformation. Armed with shovels, sheer determination and a vision for a better future, they have revived their spring following this six-step protocol. Faced with dwindling water sources and the increasing drudgery of fetching water and fodder, these women turned to a toe trench technique - simple structures designed to recharge groundwater. Their efforts have yielded good results. The spring that had been drying now flows abundantly, bringing life back to the fields. With improved water availability, fodder production has surged, saving the women time and money previously spent on procuring fodder.

Many such success stories exist in the HKH. For instance, the springs in Lohling, Bhutan, are coming back to life. Similarly, springs in Namobuddha municipality in Nepal have been revived, enhancing water security for the local population.

Call to Action

Spring revival is a strategic investment with far-reaching benefits, addressing water security; agricultural productivity; gender equality and social inclusion; and climate resilience. Reviving springs ensures reliable and cost-effective water supply for domestic, agricultural and industrial needs while promoting environmental sustainability through groundwater recharge and ecosystem restoration. It fosters economic growth by supporting agriculture, boosting eco-tourism and enabling corporate partnerships through CSR initiatives. Community-led efforts, combined with hydrogeological assessments and innovative funding models, have already demonstrated success in regions like Lohling in Bhutan, and Namobuddha municipality in Nepal as well as several locations in India. By prioritising spring rejuvenation, we can secure livelihoods, enhance climate adaptation, and promote sustainable development across water-stressed regions.

The challenge is ensuring these efforts spread across the region. Governments, researchers and communities need to work together to protect springs. This means creating policies that treat springs as critical water resources, investing in spring conservation, and promoting practices that recharge aquifers.

The Hindu Kush Himalayan region holds many lessons for the world. The drying of its springs is a reminder of how climate change and human activity can disrupt natural systems. But it is also a story of resilience and hope. By blending modern science with traditional wisdom, and by empowering local communities, especially women, we can ensure that the lifeline of the mountains keeps flowing.

Shared Responsibility

People across the Himalayan region are working hard to protect their springs. Whether it is a farmer digging trenches to recharge the soil or a scientist mapping aquifers, every effort counts. After all, water is not just a resource - it is life. For the millions of people living in the shadow of the Hindu Kush Himalayas, saving their springs means saving their future.

Moldhar's story is a microcosm of hope for the HKH. By embracing springshed management and empowering local communities - especially women, who are often the primary water managers - regions can tackle water scarcity and ensure sustainable development. The journey of Moldhar reminds us that small, grassroots actions, combined with larger policy and governance support, can restore not just water sources but livelihoods and dignity.

References

1. Bharti, N., Khandekar, N., Sengupta, P., Bhadwal, S., Kochhar, I. (2019). Dynamics of urban water supply management of two Himalayan towns in India. Water Policy (2020) 22 (S1): 65–89. https://doi.org/10.2166/wp.2019.203
2. Subramanian, S. (2024). Freshwater springs and wildlife dependence. ICIMOD Blog. Published: 12 December, 2024.
3. ICIMOD (2018). Bhutan's First Spring Revival Test Site in Lholing Village, Paro. http://www.icimod.org/?q=31936
4. ICIMOD (2020). *COVID-19 impact and policy responses in the Hindu Kush Himalaya.* International Centre for Integrated Mountain Development.
5. NITI Aayog (2018). Report of Working Group: Inventory and Revival of Springs in the Himalayas for Water Security. Retrieved from https://niti.gov.in/writereaddata/files/document_publication/doc1.pdf
6. Shrestha, RB; Desai, J; Mukherji, A; Dhakal, M; Kulkarni, H; Mahamuni, K; Bhuchar, S; Bajracharya, S (2018). Protocol for reviving springs in the Hindu Kush Himalayas: A practitioner's manual. ICIMOD Manual 2018/4. Kathmandu: ICIMOD
7. Wester, P., Mishra, A., Mukherji, A., & Shrestha, A. B. (Eds.). (2019). The Hindu Kush Himalaya Assessment: Mountains, Climate Change, Sustainability and People. Cham, Switzerland: Springer Nature.
8. Poudel, D., D., & Dues, T. W. (2017). Vanishing Springs in Nepalese Mountains Assessment of Water Sources, Farmers' Perceptions, and Climate Change Adaptation. Mountain Research and Development (MRD). 37(1) : 35-46. http://dx.doi.org/10.1659/MRD-JOURNAL-D-16-00039.1
9. Thapa, S., and Dhakal, M. (2024). Forgotten Spring Sources: Lifelines During Disasters. ICIMOD Blog. Published: 22 November, 2024.

10. Uprety, S. (2024). Rethinking Ecosystem-based Adaptation (EbA) strategies in the Indian Himalayan Region. ICIMOD Blog, Published: 23 July, 2024.
11. Thomas, S., and Kunwar, S. (2024). (Commentary) The Silent Crises of Mountain Springs and its Consequences. MONGABAY, India. Published: 22 March, 2024.
12. Lone, S., S. (2022). Panzath: Kashmir Village where a 900-year-old Spring Cleaning Festival has Conserved Traditional Bodies. 30 Stades. Published: 30 May, 2022.
13. Rafi, A., B. (2022). The Potential of Kasmir's Springs for Climate Resilience. Mongabay, India. Published: 12 September, 2022. The potential of Kashmir's springs for climate resilience
14. Census of India (2011). https://www.census2011.co.in/data/town/801634-darjiling-west-bengal.html
15. Shah, R., and Badiger, S. (2020). Conundrum or Paradox: Deconstructing the Spurious Case of Water Scarcity in the Himalayan Region Through an Institutional Economics Narrative. Water Policy 22 (2020) 146-161. https://lib.icimod.org/record/34739/files/06032020_HimalDoc_Conundrum-or-paradox-deconstructing-the-spurious.pdf
16. Bharti, N., Khandekar, N., Sengupta, P., Bhadwal, S., and Kochhar, I. (2020). Dynamics of Urban Water Supply Management of Two Himalayan Towns in India. Water Policy (2020) 22 (S1): 65–89. https://doi.org/10.2166/wp.2019.203

"We forget that the water cycle and the life cycle are one."

~Jacques Cousteau

CHAPTER 4

Frozen Treasures of the Himalaya

Arun B Shrestha

"The glaciers are the Earth's thermometers.
They don't lie, and they are telling us the planet is in a fever."
~Katharine Hayhoe

Arun B Shrestha

Dr. Arun Bhakta Shrestha is the Senior Advisor – Reducing Climate and Environmental Risks at ICIMOD. Until January 2025, he headed the Strategic Group on Reducing Climate and Environmental Risks at ICIMOD.

Dr Shrestha is recognised for his pioneering work on climate change in Nepal. He earned his PhD in Earth Sciences from the University of New Hampshire, USA. His extensive research, including field experience, spans Nepal and the Hindu Kush Himalaya. He played a pivotal role in the successful GLOF risk mitigation at Tsho Rolpa glacial lake, a first in the southern Himalayas.

He is an editor of the Hindu Kush Himalaya Assessment and a Coordinating Lead Author of its Climate Change chapter. His contributions span diverse areas related to climate, cryosphere, and water. He is also deeply involved in science-policy engagements and regional water cooperation.

Frozen Treasures of the Himalaya

Arun B Shrestha

The Hindu Kush Himalayan (HKH) region is the world's largest mountain system both vertically as well as horizontally. The HKH's climate is characterised by extreme altitudinal gradients, with high-altitude zones experiencing sub-zero temperatures year-round, sustaining its cryosphere. The climate here is shaped by two dominant moisture systems: the Southeast summer monsoon (June-September) and winter western disturbances (December-March). The summer monsoon delivers over 70% of annual precipitation to the eastern and central HKH, sourced from the Bay of Bengal and the Indian Ocean, replenishing glaciers and snowfields.[1] In contrast, winter western disturbances – moisture-laden winds from the Mediterranean – bring snowfall to the western regions.[2]

The region holds the third largest amount of snow and glacier ice outside the polar regions and is therefore often called the 'Third Pole', spanning eight countries from Afghanistan to Myanmar.[3] This vast mountainous expanse serves as the 'water tower' of South Asia, feeding ten major river systems, including the Indus, Ganges, and Brahmaputra.[4]

1. Shrestha, Agrawal et al. 2015
2. Dimri, Niyogi et al. 2015
3. Yao, Thompson et al. 2012
4. Yao, Bolch et al. 2022

The HKH's cryosphere – comprising glaciers, snow cover, and permafrost – plays a pivotal role in regulating regional water security, biodiversity, and climate systems. It's glaciers alone store approximately 12,000 cubic km of freshwater, acting as a natural buffer against seasonal water variability. The HKH is home to 240 million people, with millions more downstream reliant on its water resources. High-altitude communities, many indigenous, depend directly on cryosphere-fed springs and rivers for livelihoods, irrigation and cultural practices. The HKH cryosphere supplies water to ten major river basins of Asia, home to another 1.6 billion people living downstream. Altogether 1.9 billion people, one-third of the global population directly or indirectly rely on water and other ecosystem services from HKH.[1] However, rapid warming and socio-economic changes are destabilising these systems, exacerbating water scarcity and displacing vulnerable populations.[2]

This essay explores the frozen treasures of the Hindu Kush Himalayas (HKH) – its glaciers, snow cover and permafrost - and their crucial role in sustaining water resources, ecosystems and livelihoods. As climate change accelerates, these frozen reservoirs are shrinking, posing significant threats to water security, agriculture, energy production and increasing disaster risks. By examining the science behind the HKH cryosphere, the vulnerabilities as well as the global significance, we highlight the urgent need for action to protect and adapt to a changing cryosphere in a warming world.

Glaciers: Lifelines of the Himalayas

A glacier is a persistent body of dense ice formed over long timescales (decades to millennia) by the accumulation in high

1. Immerzeel, Lutz et al. 2019
2. Wester, Mishra et al. 2019

altitude, compaction, and recrystallisation of snow under its own weight.[1] It flows slowly downhill or outward due to gravity and survives year-round in climates where snow accumulation exceeds melt. As the glacier moves downwards, it melts due to warmer temperatures. If the accumulation is higher than melting, the glacier expands and advances, but if the melting is faster than the accumulation it shrinks and retreats.[2]

The Hindu Kush Himalayan (HKH) region encompasses approximately 54,000 glaciers, covering an area of about 60,000 square kilometres and containing an estimated 6,000 cubic kilometres of ice reserves. This accounts for roughly 1.4% of the HKH region being glaciated, with total ice reserves equal to approximately three times the annual precipitation over the area.

Glaciers in ten river basins originating from the HKH

River Basin	Glacier Number	Glacier Area (km^2)	Glacier Volume (km^3)
Amu Darya	3,277	2,566	163
Indus	18,495	21,193	2696
Ganges	7,963	9,013	793
Brahmaputra	11,497	14,021	1,303
Irrawaddy	133	35	1.29
Salween	2,113	1,352	88
Mekong	482	234.61	11
Yangtze	1,661	1,660	121
Yellow	189	137	9.24
Tarim	1,091	2,316	379
Interior	7,351	7,535	563
Total	54,252	60,063	6,128

1. NSIDC
2. Bajracharya and Shrestha 2011

Figure: Repeat photography of Rikha Samba Glacier in the Dhaulagiri region, Central Nepal, captured in: a) 2011, b) 2016, c) 2021 and d) 2024. The glacier terminus (lowest part) has retreated significantly over this period. Additionally, ice loss on both the left and right sections of the glacier has exposed underlying bedrock.[1]

Snow Cover: The Seasonal Reservoir

Snow cover refers to the layer of snow that accumulates on the Earth's surface and is measured in terms of its aerial extent, depth and duration. It plays a vital role in Earth's climate and hydrology, influencing energy balance, hydrological regimes, and ecosystems. From a hydrological perspective, the density of snow cover, which determines its water content, is a key parameter. Additionally, snow persistency – the proportion of time a specific area remains covered with snow over a given period – is an important indicator for climate, hydrology, and ecosystem dynamics. Snow cover is monitored using satellite imagery, weather stations, and remote sensing technologies.[2]

1. Source: a.-c. S.P Joshi, ICIMOD; d. T.C. Sherpa, ICIMOD
2. Hall et al., 2005

There are two types of snow cover: seasonal snow cover, which forms during a particular season and melts as temperatures rise; and permanent snow cover, which persists year-round and is typically found in polar regions and high-altitude mountains. Over time, permanent snow contributes to glacier formation.

Snow cover varies spatially and temporally, but a systematic evaluation of snow cover in the Hindu Kush Himalaya (HKH) remains limited. Existing studies estimate that the average snow-covered area in the extended HKH ranges between 10% and 18%, with seasonal variations. During winter, the maximum extent of snow cover reaches between 21% and 42%, while in summer, the minimum extent is between 2% and 4%.[1] Snow accumulation in the HKH is influenced by regional weather patterns. Western disturbances bring winter snowfall, primarily affecting the western HKH, whereas the summer monsoon contributes to snowfall at higher elevations, mainly in the eastern HKH.

Snow cover functions as a natural water reservoir, storing precipitation in winter and gradually releasing it in spring and summer, ensuring sustained river flows. On average, 24% of river flow in major HKH river basins originate from snowmelt, with notable contributions in specific basins, such as 74% in the Amu Darya Basin and 40% in the Indus Basin.[2] In most basins, snowmelt contributes more to river flows than glacier melt, playing a crucial role in agriculture, hydropower and drinking water supply. Beyond water resources, snow cover regulates ecosystems by maintaining soil moisture, supporting biodiversity and influencing vegetation cycles. Additionally, its high albedo effect helps regulate regional temperatures by reflecting solar radiation and cooling the environment.[3]

1. Immerzeel et al., 2009; Gurung et al., 2011
2. Khanal et al., 2021
3. Singh et al., 2023

Permafrost: The Hidden Frozen Ground

Permafrost refers to ground that remains frozen for at least two consecutive years. It consists of soil, rock, and organic material bound together by ice. Permafrost is widespread in the Arctic, but it is also found in high-altitude mountain regions such as the Hindu Kush Himalaya (HKH). The presence of permafrost is influenced by elevation, temperature and local climatic conditions. In the HKH, permafrost occurs primarily in high-altitude regions above 4,000 meters, where mean annual ground temperatures remain below freezing. The layer that thaws during the warm season is called the active layer.

Permafrost plays a critical role in maintaining landscape stability by binding loose sediments and preventing rockfalls, landslides and erosion. In mountain regions like the HKH, it helps stabilise steep slopes, reducing the likelihood of natural hazards. Additionally, permafrost is a major carbon store, trapping organic material that has remained frozen for thousands of years. When permafrost remains intact, it acts as a carbon sink, preventing the release of greenhouse gases into the atmosphere. Permafrost could also have an important role in hydrology and ecosystems.

Compared to the Arctic, permafrost in the HKH remains poorly studied and mapped due to the region's complex topography and limited long-term monitoring. Permafrost is estimated to cover over 2 million square km in the HKH,[1] though estimates vary considerably across studies. Direct field-based studies and high-resolution mapping are scarce resulting in high uncertainty about its thickness and response to climate change. The lack of systematic observation makes it difficult to assess permafrost stability, the role in hydrology and the risks associated with thaw.

1. Obu et al., 2019

Warming Cryosphere: Challenges and Impacts

The cryosphere in HKH is experiencing rapid changes due to climate change. These changes pose significant challenges to ecosystems, water security, infrastructure and livelihoods, both in high-altitude and downstream regions. The consequences of a warming cryosphere extend beyond these areas, influencing global climate systems, sea levels and water availability.

Rising temperatures have accelerated the retreat of glaciers, altered snowfall patterns, and increased permafrost thawing. In the Hindu Kush Himalaya (HKH) region, glaciers are losing mass at an alarming rate, 65% more in the recent decade compared to the previous one.[1]

Projections suggest a loss of up to two-thirds of their volume by the end of the century if current emissions continue.[2] Snow cover is also diminishing. Snow cover persistency has reduced in general in the HKH with 2025.[3]

The cryosphere regulates river runoff by releasing water, primarily as snowmelt during spring, and glacier melt thereafter, which also replenishes aquifers. This rapid glacier and snow melt significantly influences river flows in the region, but depends on the level of contribution of meltwater to river runoff, which varies across basins, accounting for up to 79% in the Amu Darya and as little as 5% in the Irrawaddy.

As glaciers continue to shrink, an initial increase in meltwater may lead to a temporary rise in river flows, but this is expected to decline as glaciers diminish, potentially reaching 'peak water' around mid-century. This decline poses risks to water availability for agriculture, hydropower and drinking supplies, especially during dry non-monsoon seasons, as well as in semi-arid parts of the region.

1. Muhammad 2025
2. Wester, Mishra et al. 2019, ICIMOD 2023
3. ICIMOD 2024

Declining snowfall and irregular melting patterns disrupt the natural hydrological cycle. Reduced snowfall diminishes the seasonal freshwater storage, disrupting the local water balance and impacting sectors dependent on it, such as agriculture and water supply. Conversely, erratic melting, often intensified by warm episodes, can lead to sudden and excessive runoff, heightening the risk of floods, erosions and cascading hazards like glacial lake outburst floods (GLOFs) and other hazards, similar to the 2021 Melamchi disaster.[1] The retreat of mountain glaciers has increased the size and number of glacial lakes, elevating the risk of glacial lake outburst floods (GLOFs).

There is a large uncertainty about the changes in HKH permafrost. Remote sensing, modelling and some limited ground studies suggest that permafrost in the Hindu Kush Himalaya (HKH) region is warming and degrading.[2] The degradation of permafrost also weakens soil stability, leading to increased risk of landslides and debris flow, which can damage infrastructure and settlements. This process can release previously trapped greenhouse gases such as carbon dioxide and methane, further amplifying global warming. The extent of this in HKH is yet to be assessed.

Ecosystems and biodiversity in the HKH are under threat due to cryosphere changes. The effects of the changing cryosphere on fragile mountain habitats are particularly acute, with cascading impacts reported in most ecosystems and affecting most inhabitant species. The region is already seeing both species decline and extinction.

With 67% of the HKH's ecoregions and 39% of the region's four global biodiversity hotspots outside protected areas, the HKH's extraordinary biodiversity is particularly vulnerable to climate

---

1. Maharjan, Steiner et al. 2021
2. ICIMOD 2023

impacts. Livelihoods dependent on stable cryosphere conditions, such as pastoralism and tourism, are increasingly vulnerable.

Glacial Lake Outburst Flood

A Glacial Lake Outburst Flood, or GLOF, is a sudden release of water from a lake fed by glacier melt that has formed at the side, in front, within, beneath, or on the surface of a glacier. The sudden release of water can cause devastating downstream flooding, posing significant risks to communities, infrastructure, and ecosystems.

A Glacial Lake Outburst Flood (GLOF) occurs when a glacial lake, which forms in high-altitude regions as glaciers melt, breaches its natural dam, often made of ice, debris, or moraine. This can happen due to various factors, such as rapid slope movement into the lake, heavy rainfall/snowmelt, cascading processes such as flood from a lake situated upstream, earthquakes, and melting of ice within the moraine dam weakening it and causing its failure.

A GLOF event can cause loss of life, destruction of infrastructure, ecosystem disruption and displacement of communities. The severity of impact depends on the nature of a GLOF event such as the amount of water released, duration of release, geomorphological character of the river valley and elements exposed downstream.

The Hindu Kush Himalaya (HKH) region is home to over 20,000 glacial lakes. Among them, over 200 are considered potentially dangerous due to their size, location, and downstream risk.[1] These lakes are particularly vulnerable to the effects of climate change, which is expected to increase threefold by the end of the century.[2]

1. Maharjan, Mool et al. 2018
2. Zheng, Allen et al. 2021

Caption: A typical glacial lake in Everest region, Nepal* (source: author)

Some GLOF events can have transboundary impacts. A small glacial lake, Gongbatongshaco outburst in Tibet, China in 2016 caused huge damage in the downstream Nepal part of the catchment, including severe damage to hydropower and other infrastructure.[1]

Reducing the risk and impacts of GLOFs is crucial for protecting lives, infrastructure and ecosystems. Key measures

1. Chen, Liu et al. 2023

include regular monitoring of glacial lakes using remote sensing and ground-based technologies, establishing early warning systems, and implementing risk reduction measures such as dam reinforcement or controlled drainage. Additionally, raising community awareness about GLOF risks and emergency response plans is essential to minimise loss of life and property. Given the transboundary nature of many glacial lakes, regional cooperation is vital for effective and coordinated GLOF risk management.

Why the Cryosphere Matters to Everyone

With HKH holding approximately 6,000 cubic km of glacier ice, making it the third-largest store of frozen freshwater after the Arctic and Antarctic, glacier melt from the region impacts sea-level rise. While the contribution of HKH glacier melt could be comparatively small, every cm rise is important. A cm of sea level rise is estimated to inundate an additional 10,000 square km of land and displace 1 million people.[1] This is particularly concerning for low-lying coastal areas that are already vulnerable to flooding.

Additionally, permafrost in the HKH, with an area of around 2 million square km stores vast amounts of organic carbon.[2] As temperatures rise, thawing permafrost releases carbon dioxide and methane, potent greenhouse gases that further accelerate global warming.

Snow cover also plays a role in regulating planetary temperatures through the albedo effect – it's loss means more solar radiation is absorbed, increasing regional and global temperatures.[3]

1. Li, Rowley et al. 2009
2. Obu, Westermann et al. 2019
3. Flanner, Shell et al. 2011

The HKH cryosphere is also central to the water-energy-food nexus, as it provides water for hydropower generation, irrigation and drinking supplies.[1] Reduced glacier-fed river flows could cause seasonal energy shortages, threatening economic stability in water-stressed regions such as Pakistan's Indus Basin.[2] The impact of all these can go well beyond the HKH region.

Call to Action

The rapid changes occurring in the HKH cryosphere demand urgent action at multiple levels – national, regional and global. Governments, scientific institutions and individuals must come together to mitigate these changes and adapt to their impacts. Without decisive action, the consequences will continue to intensify, threatening water security, biodiversity and livelihoods across South Asia and beyond.

Manage the Impacts of Cryosphere Change

Managing the impacts of cryosphere change is crucial to mitigate water scarcity, prevent natural disasters and protect biodiversity. Effective management strategies can help stabilise water supplies, reduce the risk of cryospheric hazards and preserve the unique ecosystems of the HKH region. For instance, implementing sustainable water management practices and enhancing disaster preparedness can significantly reduce the vulnerabilities of communities dependent on cryosphere-fed water resources.

Investing in scientific research and monitoring

There are still critical knowledge gaps requiring expanding glacier, snow and permafrost monitoring networks, modelling

1. Scott, Zhang et al. 2019
2. Immerzeel, Lutz et al. 2019

their change and impact assessments. Collaboration of scientists and civil society could be more effective and yield better results and ensure solutions are community-driven and context-specific.

Strengthening regional and global cooperation

Cooperation is essential to address cryosphere-related challenges. Countries sharing the HKH region must work together on transboundary water management, climate adaptation strategies, and early warning systems for disasters like glacial lake outburst floods (GLOFs). Global agreements such as the Paris Agreement and the United Nations Sustainable Development Goals (SDGs) provide important platforms for coordinated climate action.

Reducing greenhouse gas emissions

There is no alternative to reducing greenhouse gases to slow cryosphere degradation and its impact. Governments and industries must accelerate transitions to renewable energy, enforce policies to limit black carbon emissions and enhance carbon sequestration efforts. Lobbying for the implementation of global commitments should be strengthened. Individuals can contribute by reducing their carbon footprints through energy conservation, supporting renewable energy sources and minimising waste.

Leverage International Year for Glacier Preservation (IYGP)

IYGP and the Decade of Action for Cryospheric Sciences (2025-2034) offer opportunities to mobilise resources and expertise toward cryosphere protection. These platforms can amplify scientific research, policy engagement, and global advocacy to ensure that the HKH's frozen treasures are safeguarded for future generations.

The time to act is now. The future of the HKH cryosphere depends on collective efforts to combat climate change, implement sustainable policies and foster cooperation. By taking decisive action today, we can protect this vital region and ensure the resilience of millions who depend on its resources.

References

1. Bajracharya, S. and B. Shrestha, Eds. (2011). The status of glaciers in the Hindu Kush-Himalayan region. Kathmandu, ICIMOD.
2. Chen, N., M. Liu, S. Allen, M. Deng, N. R. Khanal, T. Peng, S. Tian, C. Huggel, K. Wu, M. Rahman and M. Somos-Valenzuela (2023). "Small outbursts into big disasters: Earthquakes exacerbate climate-driven cascade processes of the glacial lakes failure in the Himalayas." Geomorphology 422: 108539.
3. Dimri, A. P., D. Niyogi, A. P. Barros, J. Ridley, U. C. Mohanty, T. Yasunari and D. R. Sikka (2015). "Western Disturbances: A review." Reviews of Geophysics 53(2): 225-246.
4. Flanner, M. G., K. M. Shell, M. Barlage, D. K. Perovich and M. A. Tschudi (2011). "Radiative forcing and albedo feedback from the Northern Hemisphere cryosphere between 1979 and 2008." Nature Geoscience 4(3): 151-155.
5. ICIMOD (2023). Water, ice, society, and ecosystems in the Hindu Kush Himalaya: An outlook.
6. Muhammad, S. (2025). "HKH snow update 2025." ICIMOD. https://doi.org/10.53055/ICIMOD.1087
7. Immerzeel, W. W., A. F. Lutz, M. Andrade, A. Bahl, H. Biemans, T. Bolch, S. Hyde, S. Brumby, B. J. Davies, A. C. Elmore, A. Emmer, M. Feng, A. Fernandez, U. Haritashya, J. S. Kargel, M. Koppes, P. D. A. Kraaijenbrink, A. V. Kulkarni, P. A. Mayewski, S. Nepal, P. Pacheco, T. H. Painter, F. Pellicciotti, H. Rajaram, S. Rupper, A. Sinisalo, A. B. Shrestha, D. Viviroli, Y. Wada, C. Xiao, T. Yao and J. E. M. Baillie (2019). "Importance and vulnerability of the world' s water towers." Nature 577(7790): 364-369.

8. Li, X., R. Rowley, J. Kostelnick, D. Braaten, J. Meisel and K. Hulbutta (2009). "GIS analysis of global impacts from sea level rise." Photogrammetric Engineering and Remote Sensing 75: 807-818.
9. Maharjan, S., P. K. Mool, L. Wu, G. Xiao, F. Shrestha, R. B. Shrestha, N. R. Khanal, S. R. Bajracharya, S. Joshi, S. Shahi and P. Baral (2018). The Status of Glacial Lakes in the Hindu Kush Himalaya. Kathmandu, ICIMOD.
10. Maharjan, S. B., J. F. Steiner, A. B. Shrestha, A. Maharjan, S. Nepal, M. S. Shrestha, B. Bajracharya, G. Rasul, M. Shrestha, M. Jackson and N. Gupta (2021). The Melamchi flood disaster. Cascading hazard and the need for multihazard risk management. Kathmandu, ICIMOD: 20.
11. NSIDC. "Science of Glaciers." from https://nsidc.org/learn/parts-cryosphere/glaciers/science-glaciers.
12. Obu, J., S. Westermann, A. Bartsch, N. Berdnikov, H. H. Christiansen, A. Dashtseren, R. Delaloye, B. Elberling, B. Etzelmüller, A. Kholodov, A. Khomutov, A. Kääb, M. O. Leibman, A. G. Lewkowicz, S. K. Panda, V. Romanovsky, R. G. Way, A. Westergaard-Nielsen, T. Wu, J. Yamkhin and D. Zou (2019). "Northern Hemisphere permafrost map based on TTOP modelling for 2000–2016 at 1 km^2 scale." Earth-Science Reviews 193: 299-316.
13. Scott, C. A., F. Zhang, A. Mukherji, W. W. Immerzeel and D. Mustafa (2019). "Water-energy-food nexus in the Indus Basin: Realities, challenges, and opportunities." Water International 44(1): 2-13.
14. Shrestha, A. B., N. K. Agrawal, B. Alfthan, S. R. Bajracharya, J. Maréchal and B. v. Oort, Eds. (2015). The Himalayan Climate and Water Atlas: Impact of climate change on water resources in five of Asia's major river basins. Kathmandu, ICIMOD, GRID-Arendal and CICERO.
15. Wester, P., A. Mishra, A. Mukherji and A. B. Shrestha, Eds. (2019). The Hindu Kush Himalaya Assessment—Mountains, Climate

Change, Sustainability and People. Switzerland AG, Cham., Springer Nature.

16. Yao, T., T. Bolch, D. Chen, J. Gao, W. Immerzeel, S. Piao, F. Su, L. Thompson, Y. Wada, L. Wang, T. Wang, G. Wu, B. Xu, W. Yang, G. Zhang and P. Zhao (2022). "The imbalance of the Asian water tower." Nature Reviews Earth & Environment 3(10): 618-632.
17. Yao, T., L. G. Thompson, V. Mosbrugger, F. Zhang, Y. Ma, T. Luo, B. Xu, X. Yang, D. R. Joswiak, W. Wang, M. E. Joswiak, L. P. Devkota, S. Tayal, R. Jilani and R. Fayziev (2012). "Third Pole Environment (TPE)." Environmental Development 3: 52-64.
18. Zheng, G., S. K. Allen, A. Bao, J. A. Ballesteros-Cánovas, M. Huss, G. Zhang, J. Li, Y. Yuan, L. Jiang, T. Yu, W. Chen and M. Stoffel (2021). "Increasing risk of glacial lake outburst floods from future Third Pole deglaciation." Nature Climate Change 11(5): 411-417.

CHAPTER 5

Rivers at Nature's Pace

Siddharth Agarwal

"Asia's waters have long been a gauge for rulers' ambition, a yardstick of technological prowess - and a dump for the waste products of civilization. Water is, in a sense, a 'sampling device' for other sorts of change, even as changes in water ecology have had a direct effect on millions of people's lives."

~Sunil Amrith, Unruly Waters

Siddharth Agarwal

Siddharth Agarwal has been walking across India and its rivers, attempting to document and bring stories of marginalised communities and their environment into the mainstream. Since 2014, he has walked 6000+ kms, including the lengths of the rivers Ganga and Ken. He is the founder of Veditum India Foundation, where he is working to contribute to the transformation of the environmental governance landscape of India.

Rivers at Nature's Pace

Siddharth Agarwal

The processes that have formed our Earth have taken billions of years. That pace of formation, of the mixing of elements, was essential for life to arrive at where we are today. The continuous movement of water that forms the hydrological cycle of evaporation, condensation, precipitation and runoff, which helps distribute water, has its own pace, rhythm, pauses and rapids.

Rivers form a crucial component of the hydrological cycle; and much like everything else, have their own pace. This pace, however, has been disrupted by the advent of modernity. Extracting from nature for our own use is not a new phenomenon, but the disparity in the pace between nature and humans has never been so stark. There seems to be no course correction in sight in the near future.

How then do we make sense of, and understand our rivers? I found my answer in slowing down.

Slowing Down

My fascination for the Ganga sedimented over the years, having been brought up in a city on the banks of River Hooghly and crossing it often to visit my grandmother's house. This solidified further after an expedition to Gaumukh - the source of River Ganga, when I was in my 10th standard.

In 2016, I decided to walk the length of the River Ganga, from sea to source, over a target period of 4 months. By then I had already finished university and spent months cycling and walking across different parts of India.

My knowledge of rivers however was quite limited. I had chosen to walk along a river because I wanted to get disentangled from the network of roads. The river would now be my guide. I dove deep into the large troves of writing and commentary available on India's rivers.

It eventually took me six months to complete my journey from Gangasagar in West Bengal to Gangotri in Uttarakhand, walking approximately 3000 kms during this period. The journey was full of hurdles, breaks and lessons in humility.

It required a total submission to nature's intended pace for humans. The pace of walking made me form genuine connections with people, as well as with the river.

What became evident was that those, including myself, who benefit most from, and also design the transformation of our rivers into mere resource pools, do not really have a relationship with rivers or with nature. We are far removed from the implications of our choices, a distance awarded by the designs and scales of modernity. It is only when we have reached an unavoidable precipice that we begin to see the results of our choices.

The practice of walking along rivers has continued ever since my first walk, and has now transformed into a fellowship program – Moving Upstream – that I offer through my organisation, Veditum India Foundation. Along with others in the program, we have now walked the lengths of Rivers Ganga, Ken, Betwa, Sindh, Luni and parts of the Mahakali.

These walks have been individually transformative for all participants, and have caused the challenges faced by rivers as well as communities to surface to the fore. The simple act of walking has also allowed us to learn about the state of villages where highways and bridges do not pass.

Do you know what the Ganga looks like outside the popular cities of Rishikesh, Haridwar, Kanpur, Varanasi and a few others? For a country fascinated by the Ganga, we know too little, and from too few places along the river. Where do we then stand vis-a-vis other rivers? I am afraid, we are not doing too well.

This disconnection with our naturescapes has allowed destruction and mindless extraction for far too long. Dams for example have destroyed ecosystems and forcefully displaced, in almost all cases, hundreds and thousands of villages.

These practices however, are unfortunately not a thing of the past. They continue today. I will share two examples from my walks along Rivers Ganga and Ken to illustrate this point.

Between the Hills and Plains

In the mountain state of Uttarakhand, a large part of which is situated within the young Himalayan mountain ranges, the imaginations of dwellers from the plains have taken roots in the fragile hill slopes. Wide highways, no matter how ill-suited, have been the answer to all problems.

During my walk along the Ganga, the case of the Char Dham All Weather Highway Development Project in Uttarakhand was still in the Supreme Court. As the name suggests, the project intended to connect the major pilgrimage sites for easy tourist access. The fast pace of travel on wide highways in the plains is being sought in the mountains as well.

Wide highways were assumed to be all weather roads, even though their stability in these fragile mountains should have been in question. The project's proponents did not shy away from demanding clearance of forests even in the protected Bhagirathi Eco-sensitive zone, the headwaters of River Bhagirathi (Ganga).

In the process of this project's construction, I witnessed, and it is documented by others as well, that almost every environmental protection guideline was ignored. The requirement of muck dumping away from the hill slopes and river valleys, was blatantly flouted. Unethical misrepresentations were made to avoid adherence to law. The project was literally bulldozed through unstable hill slopes, further destabilising them.

The creators of 'all weather roads' were destabilising the foundations of their own roads by dumping illegally blasted mountain debris over native slope vegetation that was holding the mountain together in the first place. Not only did this destabilise the slopes, but also significantly silted the state's rivers, including the Ganga. For a state that has faced frequent large floods in recent times, these malpractices are contributing to increased mishaps and also multiply the intensity of such events when they happen.

For those of us accustomed to speeds that overlook detail, we have long ignored the natural world, and we will continue to overlook the degradation and destruction around us today.

When the Supreme Court's High Powered Committee assessed the project and offered its suggestions to build a stable road instead of a wide one, it faced stiff resistance from the bureaucrats in Delhi. The chairman, Dr. Ravi Chopra, finally submitted a minority report in his own committee, after months of diligent studies and efforts to instil sense.

The act of walking offered a pace that allowed me to witness and understand this unfolding phenomenon. It had the same effect on the few co-travellers who were walking with me along the Ganga in Uttarakhand. The residents whom we met along the way, and who powered our journey with their generosity, reminded us that the Char Dham pilgrimage is considered complete only when it is done on foot.

As the Ganga flows through Uttarakhand, bookended by a series of dams and silted by road construction debris, it is forced to slow down unnaturally where it is supposed to be fast-paced. The pacing is out of sync with the nature of the river and its natural journey.

Getting Hoodwinked at Basin Scale

"When there is excess water in the Ken, it will be transferred to the Betwa, and when there is an excess in the Betwa, it will flow to the Ken" said a few people who had heard about the Ken Betwa River Linking project.

Of all the villages we visited over a two month walk along the length of the River Ken in 2017-18, only a handful had heard about the proposed 'river-linking' of rivers Ken and Betwa. Those who had heard about the project, had mostly learnt about it through newspaper clippings that largely carried the press releases issued at regular intervals by various ministries and government departments.

When asked "How will the water which will be transported from the Ken to the Betwa through the force of gravity, make its return journey?", a lightbulb switched on for most people. They knew instinctively that no pumps would be setup for such large scale water transfers.

Built on the foundation of a name which is a misnomer, the series of proposed 'river-linking' projects are just water transfer projects, the likes of which we have seen in plenty in India and across the world. There is no 'linking' that is taking place.

A whole country is being hoodwinked by repackaging dams and hydropower projects in a shiny new outfit. This single project, in this series, will cost the tax payers over ₹ 50,000 crores. I sometimes fail to fathom the number of zeroes in there.

And this is just the basic financial cost. The project is also submerging over 6000 hectares of prime forest land inside the Panna Tiger Reserve with over 30 lakh trees, negatively impacting a Schedule 1 species that supposedly gets the highest protection under the Wildlife Protection Act. Furthermore, thousands of families will be displaced from their homes.

What makes this doubly offensive is the fact that this project has been sold to everyone as a solution for the water problems of Bundelkhand, whereas the premise of the project is to export water OUT of Bundelkhand through a water substation mechanism. Most places along the River Ken are small towns and villages, and consequently receive lower attention in the media as well as in our popular imagination.

The act of walking in this region, allowed us to learn from communities about their relationship with their river, as well as convey accurate information about this proposed project to those that will be impacted. Our media fails to do this.

A Thousand Cuts

Beyond the large-scale infrastructure and diversion projects, which sometimes get noticed because of their size, there are a thousand smaller cuts that go totally unnoticed.

In our journeys along India's rivers, we have been documenting various challenges that plague rivers and communities. A significant issue emerging from our documentation has been the rampant illegal and destructive sand mining from our riverbeds. Simply explained, sand mining refers to the extraction of sand from riverbeds, floodplains, wetlands, coasts, etc. I write particularly about the extraction from river systems.

Because this ecosystem is extremely financially rewarding, a whole syndicate of official and unofficial actors enable its continuity. They disregard established law and policy for personal and political gains, while ecological, social and fiscal losses pile up.

Again, the needs of modernity demand the sacrifice of ecosystems and communities. The vast quantities of sand required by the construction and infrastructure sectors is out of sync with the pace of a river's natural systems, of grinding down larger rocks and boulders into tiny sand particles.

What is important to remember is that not all extracted material is used for actually constructing useful infrastructure, but much of it contributes to the insatiable need of the speculative real estate market. This demand is depleting river systems at large scale, decimating whole ecosystems.

Sand mining still flies under the radar unless a mishap occurs. Compounded by the maintenance of violent spaces around sand quarries to exert control, building an understanding of this topic has traditionally required fierce and honest individuals willing to risk life and limb.

I would like to reinforce that a large number of environmental issues, including those of our rivers, continue to exist because of our collective apathy. Whether large or small issues, our disconnect with each other and with nature is at the root of the problem.

Our needs and demands will not vanish overnight, but the antidote to this may be in slowing down and being conscious of what this is doing to entire ecosystems. Perhaps we need to rebuild our connections and our communities and we need to be in consonance with nature's pace. Like a flowing river.

CHAPTER 6

Slow Water

Healthy Regulation of Earth's Ecological and Climate Systems

Alpha Lo

"Water begets water, soil is the womb, vegetation is the midwife."
~Millán Millán

Alpha Lo

Alpha Lo teaches about water in Permaculture Design Courses, and hosts and writes the Climate Water Project podcast and newsletter. He is trained in physics and works on eco and water restoration projects.

Slow Water

Healthy Regulation of Earth's Ecological and Climate Systems

Alpha Lo

Slow water is a path. Slow water is a clue, a pointer, a key, an injunction to how we can bring earth's natural systems back to equilibrium. Slow is a directive, an abbreviation of the 'slow it, sink it, spread it' maxim of how we can work with rainfall once it comes in contact with land, to help it to infiltrate into the soil. Slowing water helps set off a chain reaction of feedback loops that help regulate the earth system – the small water cycle, the large water cycle, the groundwater-surface water cycle, and the vegetation – evapotranspiration thermostat mechanism. These feedback loops hold solutions to how we can lessen droughts, fires, floods, heatwaves and water scarcity. Slowness is a way to shift away from the fast tracking of water off our continents via the dam – aqueduct-channelisation-drainage industrial complex. Slowness is a way that the terrestrial carbon and water cycles can recouple. Slow water is an anthem, a song of eco and climate restoration.

In Al Baydha, Saudia Arabia, where the vegetation had been destroyed by mismanaged grazing and firewood cutting, where rare rains come in flash floods that problematically rush off the land, villagers have been digging wide shallow swales and building granitoid-rock, traditional Incan check dams to slow the path of the rainfall on the land. The rain can then infiltrate into the land, and help grasses, mycelia, and acacia trees to once again flourish in the region.

In Niger, where the land is arid and sparse, villagers dig demi-lunes, semi-circle swales in the sloped ground, that slow and catch the rainwater. Into these they have been planting sorghum, millet and gao trees which are beginning to revegetate the region.

Each year, in many places in India, after a few months of monsoon rains the droughts come, destroying crops, and leaving villagers without a source of income. So, villagers have been digging johads (small ponds) and swales to catch the rain in the monsoon season which is then used to irrigate the crops into the dry season. The ponds also slowly seep the water into aquifers below to replenish them, helping provide more well water. Rivers have been revived as aquifers are able to supply them with water year-round. Thousands of villages have been reinvigorated as they learn the art of slowing rainwater.

On the Mississippi, a century of levees, built by locals in their area, to protect against flood, has sometimes unfortunately and counterintuitively led to the river speeding up and bigger floods happening elsewhere on the river. Now some of the levees are getting dismantled, as river engineers realise that a better way to stop floods is by slowing the water, by decelerating it in the floodplains.

In the Louisiana Mississippi basin, 17 miles of levees have been taken down to allow the water to overflow onto the floodplains, and then slow down, amidst the replanted cypress, oak and ash, and the returning waterfowl and wading birds.

Logging along the Yangtze River, and deforestation along the Nenjing and the Songhua rivers, has led to floods in

China that have affected 200 million people. With no trees and vegetated landscape upstream to slow or stop the stormwater, the downstream cities are left to absorb the vast liquid influx. To deal with the problem, the government has been reforesting river-side land, and unveiling its sponge city program in 30 pilot cities to soak and slow the stormwaters. Rivers are being recurved and revegetated, so they can slow river water. Rain gardens, road bio-swales and wetlands are being created. People and businesses are being moved off floodplains, and the land depaved, to allow the land to return to its absorbent and water-slowing nature.

Why is slowness so important? Imagine an art museum where a patron walks in every minute. If people walk briskly through the museum the total amount of people in the museum at any one time will be a lot less than if people walk slowly. If people take twenty minutes to walk through, the place will be a lot less packed than if people go slowly, taking two hours to walk through. Now let's make the analogy that the museum is the continent, and people are water. If we slow the rate of the water flow there will then be a lot more water on the continent. A key quantity here is something I call Water Path Time, which is analogous to time spent in a museum. Water Path Time is the average time for water to go into the ocean and then come back out to the ocean. If we can increase this quantity we will have a lot more water on our continents.

Let's look at California through the lens of continental water amount, and water path time. Water flows off the Californian Sierra Nevada mountain ranges, is caught by a plethora of dams, and then funnelled to Central Valley farms and Southern California cities. In Los Angeles and San Diego, the water is used and then funnelled through storm drainage and man-made infrastructures. The LA river, channelised with concrete, after the devastating 1938 floods, funnels water quickly out of the area and into the ocean. The water, as it moves from the base of the Sierra Nevadas to the Pacific, is on a fast track. It does not have the lingering time it used to. The patrons are moving too quickly through the museum. California is drying up.

We can take the museum analogy further. People in the museum can walk around twice, doubling their walk path time, resulting in a more packed museum. On a continent water can go around twice or more. It can rain down, evapotranspire back up, and then rain down again, in a process known as the small water cycle. The more the water cycles the more water there is on the continent, just like how the museum will have more people inside the building if more people walk around again. Globally about 65% of our rain comes from the small water cycle, and 35% comes from the large water cycle. The large water cycle is where the ocean vapor blows inland, falls as rain, and then goes back out to the ocean. The amount of water in the small water cycle depends on how we tend to the land. If the land is able to slow the water, and guide it into the vegetation and soil to evapotranspire, then the small water cycle increases, and the amount of water on our continents increases. The small water cycle is nature's act of recursive genius in terrestrial hydration.

There is also another way for the water to go around in circles. It can sink down into the aquifers, where tree roots bring the water up, where the mycelia then spread the water around the soil, in a process called hydraulic redistribution. In the Amazon, tree roots bring the water down low during wet season, then they bring the water back up during the dry season. This is the groundwater-surface water cycle. It is as if there was a basement in the museum, where patrons could go chill for a while, and then come back up to continue looking at the art.

All over the world there are water scarcity crises, tense water debates, and water-induced political turmoil. Water has been re-allocated, and overallocated; local and national governments are scrambling to make sure their constituents can get water. Increasing the amount of water on the land would help with these issues significantly. To do that we need to slow the movement of the patrons through the museum, and allow them to go around multiple times.

Before the onslaught of dams, aqueducts, pavement, and farm drainage, rainwater would form wetlands all over the Central Valley and Los Angeles, and get absorbed in greater quantities into the Californian wilderness. There was thus more water on the land available to evapotranspire and get blown towards the Sierra Nevadas, where it then came down as rain and snow. The amount of water in the small water cycle was larger. It is the equivalent of patrons walking a second time again around the museum.

As a result of all this loss of continental water, there are many more water shortages in California. The wilderness is not as lush and hydrated. Less wetlands means the winds are hotter and drier, creating more wildfire. Less wetlands also means less groundwater. It is groundwater that supplies forty percent of water to the states. Rivers, which depend on aquifers to refill, no longer flow year-round. Rain is less as small water cycles decrease.

There are a few signs that some Californians are beginning to understand the water cycle, and to hear the rhythms of this slow water anthem. A Sustainable Groundwater Management Act has been passed. Los Angeles is starting to regreen its river, unstraighten its length, so the waters flowing down can help grow the vegetation instead of just rushing out to sea. In Burbank, plans for the Rory Shaw wetlands is unfolding to absorb stormwaters from around the area, cleanse it, and then sink it to aquifers below, where wells can then bring the water up to LA's residents when needed. LA is working to create a human assisted groundwater-surface water cycle. Half of LA's residents could get enough water just from rain, if the rain was recycled in these wetland-aquifer systems. If LA can supply its own water, it would not have to be depleted from mountains up north. Other wetlands projects are also in the works. In the Central Valley they are refilling aquifers by creating flooding farms and creating temporary wetlands on them. And there is a small movement to get industrial farms to convert to regenerative agriculture so they can better slow the water.

Slowing water also helps the large water cycle, which is where ocean vapor blows inland, falls as rain, and then flows back out along rivers to the sea. We need the rivers to be flowing well into the ocean year-round, otherwise salt water will intrude into the continent. If water is guided into the land more, it would flow more slowly underground, and be able to come out in the dry season to keep the rivers flowing. Many of the USA rivers are channelised and so the water rushes off them in the wet season, and does not get a chance to stay year-round.

Allowing rivers to meander again, to form oxbows and curves, helping willows and other trees regrow which slow the waters, reintroducing beavers so they help the rivers overflow laterally into adjacent wetlands, will slow the water so that it can continue moving out all year. It is as if we wanted people coming out of the museum 24 hours a day, and yet patrons only entered during the day.

Another reason it is so important to slow water, thereby increasing the amount of water on the continent, is because plants regulate the heat of the land through water. As it gets hotter the trees and vegetation evapotranspire more water to cool the land, similar to how we sweat to cool. The continent is like a large swamp cooler, which is a simple low tech air conditioner which cools the air by evaporating the water in its container. If the swamp cooler does not have enough water, then it can only cool your room that much. If the continent does not have enough water the ecosystem can cool the continent during heatwaves. We have to allow the natural ecosystem to remain in its natural state to be able to cool the land locally. We have to slow the water to cool the land. Slow water cools.

Forests evapotranspire so much they often create low hanging clouds. These clouds shield the earth from heat. The heat and water equation is complex. Water is also a greenhouse gas, and so can also heat. Research is appearing recently that suggests that the overall effect on the earth from the forest evapotranspiration is a net cooling.

A key actor, perhaps the major actor in the art of slowing water is our soil. As microbes, fungi, and organic matter multiply they are able to create air pores in the soil, which then allow the soil to absorb a lot more water. For each one percent increase in organic matter in the soil, the soil will be able to absorb 20,000 gallons more water per acre.

Industrial agriculture and its use of pesticides kill the insects that help make soil by decomposing organic matter. The use of synthetic fertiliser destroys the natural food webs in the soil. Tilling breaks up the mycelia strands, and monocropping destroys the biodiversity. These practices have degraded the soil gradually over decades. When the rains come, or when water is piped in from aqueducts it does not stay on the land long, but is instead piped to rivers and off the continent. The soil degradation has been increasing the velocity with which water leaves the continent.

A regenerative agriculture movement, with its permaculturists, agroforesters, syntropic agriculturists, silvo-pasturists, biodynamicists, natural sequence farmers, and indigenous practitioners has been working to restore the soil. This is done by restoring the soil carbon sponge, to restore the microbes and fungi to the soil. We need to understand how ecological succession leads to more biodiversity which then builds a more complex soil web, as more complex carbon structures are generated, and then also decomposed. It's a cycle, a natural engine that gets more and more efficient at building soil, and slowing water.

There is some subtlety here about the axiom of slowness in regards to large man-made dams. Dams ostensibly slow water at first glance. However, there are some caveats to this. On many lands, extreme rainfalls are coming more frequently. When they approach, many dams have to release their water, so they have an appropriate buffer for the large storms. That water then rushes out at high velocity along the river to the ocean. It is as if museum patrons were corralled into one room, and then all of sudden were escorted quickly to the exit. In addition, a lot of water in today's dams are being funnelled to cities which use the water once and then flush it out to the ocean, so that water is fast tracked out. In this case the analogy, is the patrons are kept in one room and then guided to see one painting and then escorted out of the museum. At this stage of scientific progress, it is easier to get our water to do one thing, and forget the large picture, that the water is playing many roles all at once. In trying to engineer our water to do one thing, we throw the whole system out of balance.

Dams are also not in tune with the environment. If the dam water was instead distributed amongst many wetlands, they would be growing vegetation, aiding the microbial, insect and wildlife around the wetlands, and naturally cleansing the water.

Dams also change the temperature of the water, and block the flow of fish, like salmon which could otherwise swim upstream, where it then additionally becomes important for the ecosystem by being part of the food chain of bears and birds. Mussels, which cleanse the water, are also inhibited in their growth. Moreover, dams block the flow of sediment downstream, where it could have turned into soil which helps the ecosystem flourish.

Complex systems synchronise through rhythms. If you change that, you can throw off the whole system. If ones heart beats too fast then you start to get lightheaded, it causes shortness of breath, and the organs and tissues may not get enough oxygen. So it is with the geophysiology of the earth. The earth is one interconnected self-regulating system. The water is part of its circulatory system. If it has an innate rhythm, an innate velocity, that helps it synchronise with biology, geology, and climatology.

The planet is a self-regulating system with many feedback loops, in a similar way to how the human body regulates itself through its endocrine, neuro, and immuno systems with a multitude of feedback loops. The Gaia hypothesis, first espoused by James Lovelock in the 1970s and subsequently developed by many others, suggested that biology on earth regulated the environment, and created climatological and biochemical conditions that are amenable to life. Organisms created the oxygen in our atmosphere, vegetation evolved in ways that reflected the influence of albedo on climate.

Water forms key feedback loops in the terrestrial regulatory system - the small water cycle, the groundwater-surface water circle, the large water cycle, and vegetation-evapotranspiration thermostat. All these depend on the slowing of water. The biology, the microbiomes, the fungi, the vegetation, and the fauna can regulate the climate, in a Gaian way, through these feedback loops. If these regulatory systems get weakened, then we get more extreme weather, more drought, more large storms, more floods, and more heatwaves.

It is time for a new and yet very ancient song, the slow water song. The earth's water has its rhythm, its natural speed. Millions of years of evolution has guided the terrestrial environment and atmosphere to harmonize its water dance. And this water dance happens at a slow tempo, a slower vibration than our out-of-control modern speed obsessed culture. With the slowing of water, our cities will be relieved of constant water scarcity concerns and water cleanliness issues. Our ability to lessen droughts, floods, fires, and heat waves increases. It's time for slow water legislation, slow water practitioners, slow water engineers, slow water urban designers to join the indigenous and regenerative agriculturists and permaculturists.

It is time for the slow water anthem.

CHAPTER 7

The Hidden Dance of Rainfall

Ali Bin Shahid

”.برای باران از آسمان التماس مکن؛ خاک باش که او را می‌خواند“

*(Barā-ye bārān, be āsmān rū makon; khāk sho,
tā bārān khod forūd āyad)*

*Do not look to the sky for rain; become the earth,
and the rain will descend on its own.*

Ali Bin Shahid

Ali Bin Shahid is a systems architect based in Islamabad, Pakistan, working at the intersection of climate repair, ecological intelligence and spiritual coherence. With a background in mechatronics engineering and training in regenerative land design, Ali blends the precision of engineering with insights from prophetic medicine, hydrology and indigenous systems. He has studied under Masanobu Fukuoka's lineage and follows the Regrarians framework for land regeneration.

His work centres on repairing the small water cycle and designing landscapes that speak fluently with the sky. This includes building models that help understand how vegetation, soil and terrain influence rainfall, humidity and temperature. His approach draws from both scientific and scriptural roots, including deep study in traditional sciences, and a Qur'an-based understanding of natural systems — where water, wind, soil and sky follow precise laws.

Ali also authors the Substack blog Regenesis (r3genesis.substack.com), where he shares practical insights for planetary healing. His writing weaves together atmospheric physics, land design and spiritual pattern recognition to reimagine the human role in Earth's repair.

The Hidden Dance of Rainfall

Simple Laws That Govern Complex Weather

Ali Bin Shahid

Why do some places get plenty of rain while their neighbours stay thirsty?

You are driving through the countryside and notice something peculiar. On one side of the road is a lush green valley filled with trees, where morning mist clings to the hillsides and the air feels soft and humid. Just a few miles away, on the other side of a hill, the landscape is brown and parched, with sparse vegetation and dust swirling in the dry air. Both places have the same weather systems, are under the same sky, and yet one thrives with abundant rainfall while the other struggles with drought. It may not always be so drastic, but we are now all familiar with this phenomenon.

It is happening everywhere, all the time. Green valleys and brown hills, entire countries that seem blessed or cursed by their rainfall patterns. Why does nature seem to play favourites?

The answer is not mysterious weather gods or random atmospheric chaos. Rainfall follows simple observable laws, principles so straightforward that once you understand them, you will begin to notice them everywhere. More importantly, you will realise that we are not passive victims of weather patterns. We are active participants in a dance between land and sky. When we know the steps, we can help choreograph the outcome.

This dance has been happening for millions of years, but we have only recently begun to understand its rules. What we have discovered is revolutionary: rainfall is not just something that happens to the landscape, it is something the landscape actively creates. Every plant, every patch of soil, every hill and valley is constantly in conversation with the atmosphere above, influencing whether the sky will be generous with its moisture or hold it back.

Which reminds me of a beautiful tradition:

"When people become corrupt, Allah withholds rain from the sky. Were it not for the animals, no rain would fall upon them."

"إذا فسد الناس حبس الله عنهم القطر، ولولا البهائم لم يُمطَروا"[1]

The immediate connection we make is moral: that human injustice leads to divine withholding. But the word used here, *fasād* means more than just 'sin'. It means to distort, to unbalance, to break the order that sustains life. And rainfall is one of the most sensitive indicators of that balance.

Allah does not just speak of rain as mercy, but as a finely-tuned process, and with remarkable clarity it is said in *Surah An-Nūr (24:43):*

"Do you not see that Allah drives the clouds, then joins them together, then makes them layered and you see the rain emerge from their midst..."

أَلَمْ تَرَ أَنَّ ٱللَّهَ يُزْجِى سَحَابًا ثُمَّ يُؤَلِّفُ بَيْنَهُۥ ثُمَّ يَجْعَلُهُۥ رُكَامًا فَتَرَى ٱلْوَدْقَ يَخْرُجُ مِنْ خِلَـٰلِهِۦ

Each verb: *yuzjī* (He drives), *yu'allif* (He joins), *yaj'alu* (He layers), is a step in a design. Not random, not chaotic. This is order made visible, a choreography of atmosphere and mercy.

1. Reported in al-Bayhaqi's *Shu'ab al-Īmān*, with parallel meanings in the Musnad of Imam Ahmad

When fasād takes root, whether through greed, ecological abuse or moral injustice, we are not just spiritually damaging ourselves. We are interfering with a system that includes clouds, soil, plants, animals and air. We are interrupting the sky's response to the land.

And this is where revelation and science begin to meet.

What modern science calls 'biogenic aerosols' and 'evapotranspiration', this prophetic tradition simply called *rahmah* (mercy). It understood that animals and plants serve as intermediaries, preserving the life-giving circuit of rainfall even when human actions would otherwise break it. The implications are profound. If landscapes can create rainfall, then by healing the land, we can help heal the sky.

By understanding how this dance works, we can begin to transform drought-prone areas into rain-blessed regions, not with expensive technology or climate manipulation, but by cooperating with nature's own design.

Let us start by understanding the three fundamental principles that govern this hidden dance of rainfall.

Part I: The Three Universal Principles

Principle #1: The Sky Mirrors the Land

As is the land so is the sky. Barren land creates barren sky, while living thriving land, is reflected in a living sky. Think of the land as constantly *breathing* with the sky. Just as your breath fogs a mirror, showing what is happening inside your body, the sky reflects what is happening on the ground below. This is not poetry, it is physics in action, happening every moment of every day.

When you walk from a parking lot into a forest on a hot summer day, you immediately feel the difference. The asphalt radiates heat that makes the air restless and chaotic, creating those shimmering waves you see rising from a hot pavement. Step into the forest, and the air becomes soft and stable. The trees are working like nature's air conditioning system, not just providing shade, but actively cooling and humidifying the air around.

This difference does not merely stay at ground level, it travels upward, affecting cloud formation thousands of feet above. The chaotic hot air from degraded landscapes creates chaotic weather: sudden downpours that cause flooding, followed by long dry spells. The stable, moist air from healthy landscapes creates stable weather: gentle, regular rainfall that soaks in rather than running off.

This principle can be observed in action almost anywhere. Notice how storms often begin over forests and mountains rather than over cities or bare fields. Watch how morning mist gathers around trees but rarely over concrete. Pay attention to how differently the air feels over a lush garden versus a gravel parking lot. The land is literally breathing different types of air into the sky, and the sky responds accordingly.

This relationship is so consistent that scientists can actually predict rainfall patterns by looking at land use maps. Areas with healthy vegetation get more reliable precipitation, while degraded areas experience more extreme weather swings. It is as if the land teaches the sky how to behave.

Principle #2: Water Connects the Earth to the Sky

The second principle reveals that water creates an endless conversation between land and sky. Plants act like nature's air conditioners, constantly adding moisture to the atmosphere through an elegant process that connects soil to clouds in an unbroken cycle.

Here is how this invisible river system works: roots draw water from soil, stems transport it upward, and leaves release it as invisible vapour that rises into the atmosphere. This is not just plant biology, it is climate engineering.[1] Every leaf is a tiny humidity generator; every tree a moisture pump that helps feed the clouds above.

This principle can be experienced directly. On a summer morning, walk barefoot from your lawn onto a concrete driveway. Feel the difference beneath your feet. The grass feels cool and refreshing because it is actively pumping moisture into the air, creating its own microclimate. The concrete feels hot and harsh because it is a dead surface that only absorbs and radiates heat.

This difference scales up dramatically. A single large tree can release hundreds of gallons of water into the atmosphere each day. A forest pumps so much moisture into the air that it creates its own weather systems. Scientists have discovered that the Amazon rainforest actually creates much of its own rainfall through this process, the trees do not just depend on rain, they actively make it.

The beautiful thing about this system is that it is self-reinforcing. More vegetation means more moisture in the air, which means more clouds, which means more rain, which means more vegetation can grow. It is nature's own amplification system, turning small changes into big improvements over time.

But the system works in reverse too. Remove the vegetation and you break the moisture cycle. Less vegetation means less humidity, fewer clouds, less rain, and therefore less vegetation. This is how healthy landscapes can flip into degraded ones, and why some regions seem trapped in drought cycles.

1. Trenberth et al., 2007

Principle #3: Water Needs Time to Work Magic

The third principle might be the most important for anyone wanting to work with these systems: fast water destroys, slow water creates life.

When water moves too quickly across the landscape, rushing off rooftops, racing down streets, flooding through degraded soil, it becomes destructive. It carries away topsoil, causes erosion and disappears before it creates benefit. When water moves slowly, soaking into healthy soil, flowing gently through vegetation, pooling temporarily in natural depressions, it becomes the foundation of life.

Nature has a simple rule for managing water: 'Slow it down, spread it out, soak it in'. Everywhere you look at healthy landscapes, you will see systems designed to do this. Forest floors covered in leaves that cushion rainfall and help it infiltrate slowly. Meandering streams that spread water across floodplains. Root systems that create countless tiny channels for water to penetrate deep into soil.

You can see this principle in your own yard. After a heavy rain, notice how water behaves on different surfaces. On concrete or compacted soil, it sheets off quickly, often causing erosion. But in areas with good vegetation and loose soil, the same amount of rain soaks in gently, feeding plants and replenishing groundwater.

That is why cities flood while forests quietly absorb the same rain. Amount of rainfall matters less than how the land handles it. Fast landscapes create flood-drought cycles: too much water when it rains, not enough water between rains. Slow landscapes create stability: steady infiltration during rains, sustained moisture between rains.

The magic happens when water has time to interact with living systems. Slow water feeds soil organisms, supports root systems, recharges aquifers, and maintains the moisture that keeps the land-sky conversation going. Fast water does none of these things, it creates problems and disappears.

Understanding this principle is crucial because it reveals that managing rainfall is not merely about getting more water from the sky. It is about creating landscapes that can work with whatever water they receive, turning every drop into an investment in future resilience rather than a source of temporary chaos.

Part II: The Laws of Rainfall - Nature's Simple Rules

Now that we understand the basic principles, let us dive deeper into the specific laws that govern how rainfall actually forms. These are not complex meteorological theories, they are simple rules you can observe in nature, rules that explain why some places seem to attract rain while others repel it.

Law #1: The Dance Partner Law

Why do clouds need something to grab onto? What makes rain actually fall? Here is something that might surprise you: the air around you right now is probably full of invisible water vapour, but that vapour cannot become rain on its own. Water vapour is like invisible dancers floating around in the sky, looking for partners to dance with. Without those partners, the dancers just keep floating, no matter how much moisture is in the air.

So what are these mysterious dance partners? They are tiny particles that plants release into the atmosphere: microscopic bits of pollen, fungal spores, bacteria and other organic compounds.[1] Scientists call them 'cloud condensation

1. Carslaw et al., 2010

nuclei', but think of them as nature's way of giving water vapour something to grab onto.

Here is the beautiful part: plants are constantly releasing these microscopic dance partners. Every time you walk through a garden and smell that fresh green fragrance, you are actually breathing in some of these particles. Every time you see pollen dusting your car, you are seeing visible evidence of the invisible cloud-making material that plants pump into the air.

Next time you are out early in the morning, notice how mist and fog tend to form around trees and vegetation, but rarely over concrete or asphalt. That is the dance partner law in action. The trees are providing millions of tiny nuclei for water vapour to condense around, while the concrete offers nothing for the moisture to grab onto.

The math is simple: more plants = more dance partners = more opportunities for water vapour to condense = more clouds = more rain.

This is why forested areas consistently receive more precipitation than adjacent barren areas, even when they are experiencing identical weather systems. But here is what makes this law really powerful: different plants produce different types and amounts of these particles. Some trees, like pines and eucalyptus, are absolute champions at creating cloud condensation nuclei. They are like having a really good DJ at the dance party, suddenly everyone wants to participate.

This explains something you might have noticed. Certain types of landscapes seem to attract more morning dew, more mist, more of that soft, humid air that feels so different from dry, sterile environments. It is not just that these places are cooler or more sheltered, they are actively creating the conditions for moisture to condense from the air.

Law #2: The Goldilocks Law

Why are some places just right for rain? What makes air want to rise and cool down? Air is fundamentally lazy. It will only rise up in the atmosphere if conditions are just right, not too hot, not too cool, but somewhere in the perfect middle zone that meteorologists call 'just right for convection'.

Think of air as though you are trying to get out of bed on a weekend morning. If the room is too cold, you stay snuggled under the covers (the air stays down). If it is uncomfortably hot, you might thrash around restlessly but not really get up productively (chaotic, destructive weather). But when the temperature is just right, you naturally rise and start your day in a calm, productive way.

Plants are nature's thermostats for creating these 'just right' conditions. Through their constant release of water vapour, they moderate temperatures and create the stable conditions that encourage air to rise gently and consistently. Too hot leads to violent, chaotic updrafts that create severe storms but little useful rainfall. Too cool leads to stagnant air that never rises enough to form rain clouds.

Pay attention to how different the air feels in different environments throughout the day. In a forest or well-vegetated area, temperatures feel more stable and comfortable, even during extreme weather. The air feels 'soft' rather than harsh. Over concrete or bare soil, temperatures swing more dramatically, and the air often feels either oppressively hot or uncomfortably dry.

This is why some landscapes seem to have naturally pleasant weather while others experience extremes. The vegetation is literally moderating the atmospheric conditions, keeping things in that sweet spot where gentle, productive rainfall is most likely to occur.

Law #3: The Breathing Room Law

How do land and sky talk to each other? Why does the weather change based on what is on the ground? The bottom layer of the atmosphere, roughly the first mile above ground, acts like the land's 'breathing space'. Everything that happens on the ground directly influences what happens in this layer, and everything that happens in this layer directly affects cloud formation above.[1]

Think of this breathing space as a translation zone where the land's 'language' gets converted into sky 'language'. Healthy land speaks in cool, moist, stable tones that the sky interprets as 'let's make some gentle rain'. Degraded land speaks in hot, dry, chaotic tones that the sky interprets as 'let's either make a violent storm or no rain at all'.

Concrete and asphalt heat up rapidly and create restless, turbulent air that rises chaotically. This kind of 'shouting' from the land often leads to either no response from the sky, or sudden, violent responses like thunderstorms that dump too much water too quickly and then disappear.

Plants, by contrast, create calm, steady 'conversations' with the sky. Through their constant cooling and humidifying effects, they help maintain stable conditions in this breathing space that encourage steady, reliable weather patterns.

Notice how storms often begin over forests and mountains rather than over cities or parking lots. This is not coincidence, it is the breathing room law in action. The stable, moist air over vegetation provides better conditions for organised cloud development, while the chaotic air over developed areas either prevents cloud formation or creates the kind of unstable conditions that lead to severe weather.

1. (Stull, 1988)

Law #4: The Mountain Magic Law

Why hills determine weather. How does height change everything? Air has to work to climb hills, just like you do when you are hiking. And just like you get tired and need to rest when you have been climbing, air gets 'tired' as it rises and needs to drop some of its load. The load that air carries is moisture. As air climbs up a hillside, it expands and cools down because there is less atmospheric pressure at higher elevations. As it cools, it cannot hold as much moisture, so it starts dropping water as clouds and eventually as rain.[1]

But here is the crucial detail: the gentler the slope, the more reliable and useful the rain. Steep mountains can create dramatic rainfall, but it often comes down too hard and fast to be useful. Gentle hills and gradual elevation changes create steady, soaking rains that landscapes can actually use.

If you live where there is varied elevation, you may have noticed that mountain communities often have very different weather from nearby valleys, even when they are just a few miles apart. The higher you go, the more likely you are to encounter clouds, mist and precipitation. This is the mountain magic law working at a small scale.

This principle explains why strategic placement of vegetation on hillsides can dramatically increase local rainfall. By creating the right combination of elevation change and moisture sources (through evapotranspiration), you can essentially create artificial 'mountains' that encourage precipitation.

Law #5: The Wind Whisper Law

The difference between breezes that help vs. hurt. How does moving air affect rain formation? Wind can be either rainfall's best friend or its worst enemy, depending on how it behaves. Gentle,

1. (Whiteman, 2000)

steady winds act like conveyor belts, carrying moisture from one place to another and delivering it where it is needed. But rough, turbulent winds are like bullies that tear apart clouds before they can organise themselves into rain-making systems.

The difference lies in what the wind encounters as it moves across the landscape. When wind flows over smooth, vegetated surfaces, it tends to remain laminar, flowing in steady, organised patterns. When it hits obstacles, rough surfaces or dramatic temperature differences, it becomes turbulent.

Trees and well-planned vegetation can actually train winds to behave better. Windbreaks do not just block wind, they smooth it out, reducing turbulence and creating more organised airflow patterns that support rather than disrupt cloud formation.

Pay attention to how different areas feel on windy days. Some places feel gusty and chaotic, while others feel breezy but calm. Sheltered valleys and well-forested areas often experience gentler winds that still provide air movement without the harshness. These are the places where wind is helping rather than hindering rainfall formation.

This law explains why some regions experience reliable precipitation even in windy conditions, while others struggle with drought despite having winds that carry moisture overhead. It is not just about the presence of wind or moisture, it is about whether the landscape helps organise these elements into productive weather patterns.

Understanding these five laws gives us a complete picture of how rainfall really works. It is not random or mysterious, it is the predictable result of how land and sky interact through these simple, observable principles. And once you understand these laws, you can start to see opportunities everywhere to work with them rather than against them.

Part III: Proof These Laws Work

But do these laws actually work in the real world? Can understanding them really help us transform drought-prone landscapes into rain-blessed regions? The answer comes from one of the most remarkable climate success stories of our time: the transformation of Brazil's Cerrado region.

1. How Brazil Turned Grassland into a Rainmaker

When Simple Laws Create Big Changes

The Brazilian Cerrado used to be what scientists politely called 'degraded grassland', thousands of square miles of sparse vegetation that received barely enough rainfall to support even hardy grasses. It was the kind of landscape that seemed permanently stuck in a drought cycle: too little rain to support much vegetation, and too little vegetation to help create more rain.

But Brazilian scientists and land managers decided to test whether the laws of rainfall could work at a massive scale. They did not use expensive technology or attempt to directly manipulate weather systems. Instead, they systematically applied the five laws we just learned about.

First, they addressed the dance partner law by planting forests filled with trees that excel at producing cloud condensation nuclei. They chose species that pump enormous amounts of those microscopic 'dance invitations' into the atmosphere, giving water vapour plenty of partners to condense around.

Next, they worked with the Goldilocks law by creating the stable temperature conditions that encourage gentle, productive rainfall rather than chaotic weather extremes. Instead of the hot, dry grassland that created unstable air masses, they established vegetation that naturally moderates temperature and humidity.

They applied the breathing room law by ensuring that their new forests would create calm, stable conditions in that crucial bottom layer of the atmosphere. Rather than the restless air that rises chaotically from bare ground, their forests would breathe cool, moist air that encourages organised cloud formation.

The mountain magic law came into play through careful attention to topography, using natural elevation changes to enhance the rainfall-generating effects of their new forests.

And they managed the wind whisper law by creating forest patterns that would smooth and organise airflow rather than creating turbulence that tears apart developing weather systems.

The result? Nature responded exactly as the laws predicted. The region now receives an additional 6 inches of rain every year.[1] That might not sound like much, but it represents the difference between a landscape that struggles to survive and one that thrives.

But the really remarkable thing is that this was not a one-time improvement. Each year, as the forests matured and expanded, the rainfall became more reliable and the effects more pronounced. The land began teaching the sky new patterns, and the sky began responding with increasingly generous precipitation. Which led scientists to ask: *How much is enough to shift the weather?*

2. The Magic Number Discovery

How Much Land Do You Need to Change Weather?

One of the most important discoveries from the Cerrado project was what researchers call the 'magic number', the minimum amount of forest needed to actually influence regional weather patterns.

It turns out that small patches of forest, no matter how well-designed, do not create measurable changes in rainfall.

1. Oliveira et al., 2014

The atmosphere is simply too big to be influenced by small interventions. But there is a threshold point where forests become large enough to start conversations with the sky which cause the sky to notice and respond.

That magic number is approximately 12 square miles of continuous forest.[1] Smaller than that, and you are essentially having a whispered conversation in a noisy room, the atmosphere does not really hear you. Larger than that, and your forests begin creating their own weather systems that can influence precipitation patterns across much broader areas.

This discovery was crucial because it revealed that effective rainfall enhancement is not about having perfect forests everywhere, it is about creating strategic concentrations of forest that are large enough to activate these natural laws. Think of it like needing enough musicians to make an orchestra. A few scattered instruments cannot create symphonic music, but once you reach a critical mass, something magical happens.

The Cerrado project found that connected forests are even more powerful than isolated ones. When multiple 12-square-mile forest patches are linked by corridors of vegetation, they begin working together, creating rainfall enhancement effects that extend far beyond their own boundaries. It is like having multiple orchestras that can hear each other and start playing in harmony.

This principle explains why some restoration efforts succeed dramatically while others show minimal results. The successful ones reach this magic threshold where natural laws can take over and amplify human efforts. The unsuccessful ones, no matter how well-intentioned, simply are not large enough to trigger these atmospheric responses.

---

1. Makarieva and Gorshkov, 2007

3. Recipes for Different Places

How to Apply These Laws Anywhere

The beauty of understanding these laws is that they work everywhere, but the specific 'recipes' for applying them vary depending on local conditions. The Cerrado success story provides a template that can be adapted to virtually any climate.

- **Hot dry places** need to focus on what researchers call 'water-smart' plants, species that can survive with limited rainfall but still excel at producing cloud condensation nuclei. The strategy here is to establish hardy forests that can bootstrap the system, gradually creating more favourable conditions that allow for even more vegetation over time. Desert regions around the world have begun using this approach, starting with drought-tolerant trees that nevertheless pump aerosols into the atmosphere.
- **Coastal areas** have the advantage of abundant atmospheric moisture, so the focus shifts to capturing and organising that moisture using elevation and strategic forest placement.[1] The mountain magic law becomes particularly important in these regions, where even modest hills covered with the right vegetation can dramatically increase local precipitation. Mediterranean regions have seen remarkable success with this approach, using coastal forests and terraced hillsides to capture maritime moisture.
- **Big continental areas** like the Cerrado require thinking at the largest scales, creating forest networks that can generate their own weather systems rather than just

1. Giannakopoulos et al., 2010

capturing existing moisture. This is where the magic number principle becomes most important, because continental interiors often need multiple connected forest systems working together to create reliable precipitation patterns.

In each case, the laws remain the same, but the emphasis shifts. Arid regions focus heavily on the dance partner law, making sure every bit of atmospheric moisture has plenty of opportunities to condense. Coastal regions emphasize the mountain magic law, using topography to their advantage. Continental regions prioritise reaching the magic number threshold where forests become powerful enough to create their own weather.

What is remarkable is how consistent the results have been across different climates and continents. When these laws are applied systematically at appropriate scales, landscapes that have struggled with unreliable rainfall begin developing more stable precipitation patterns within just a few years.

The Cerrado story is not unique, it is simply a well-documented example of what happens when we work with natural laws rather than against them. Similar transformations are happening in degraded landscapes around the world, proving that these principles are not limited to any particular region or climate type.

Understanding these laws and seeing them work at scale transforms how we think about rainfall and drought. Instead of seeing weather as something that happens to us, we begin to see it as something we can participate in creating. Instead of feeling helpless in the face of changing precipitation patterns, we can start designing landscapes that actively generate the conditions for abundance.

Part IV: Your Practical Playbook

Now comes the exciting part: taking these laws and putting them to work in your own landscape. Whether you are managing a backyard garden, a farm, or thinking about larger community projects, these principles scale from the smallest to the largest applications. The key is learning to read your landscape like a detective, understanding what story it is currently telling the sky, and then gradually helping it tell a better story.

Reading Your Landscape Like a Weather Detective

What Clues Tell You Where Rain Wants to Happen? Before you start changing anything, spend time becoming a weather detective in your own area. The landscape is constantly giving you clues about how it currently interacts with the atmosphere. Learning to read these clues will guide every decision you make.

Clue #1: Follow the Water

Start with the water story. Where does water naturally want to flow when it rains? Do not just look at dramatic events, pay attention to where moisture lingers after light rains, where morning dew collects most heavily, where the air feels different on humid days. These are the places where your landscape is already having positive conversations with the sky. These are your foundation areas, the places to build from.

Clue #2: Map the Dance Partners

Next, investigate your current 'dance partner' situation. Walk around your property or community during different seasons and pay attention to plant diversity and density. Where do you see the most robust vegetation? Where does the air feel most humid and comfortable? These areas are already producing cloud condensation nuclei and moderating local climate. Notice also where vegetation struggles or where you get those harsh, dry microclimates; these are your opportunity zones.

Clue #3: Elevation is Leverage

Look for elevation changes, even subtle ones. Remember the mountain magic law: even small hills can make a difference if you work with them correctly. A slope of just a few feet can become significant if you combine it with the right vegetation strategy.

Clue #4: Listen to the Wind

Pay attention to wind patterns throughout different seasons. Where does wind feel gentle and organised versus gusty and chaotic? Where are the naturally sheltered areas that could serve as nurseries for establishing new vegetation? Understanding your wind patterns will help you design plantings that smooth and organise airflow rather than creating more turbulence.

Clue #5: Watch How Water Moves

Observe how your landscape currently handles the 'slow it, spread it, sink it' principle. Where does water naturally slow down and infiltrate? Where does it rush off quickly? The areas where water already moves slowly are your natural allies, places where the landscape is already demonstrating good water behaviour that you can enhance and expand.

Starting Your Own Rain Project

How Do You Begin Changing Weather?

The beautiful thing about working with these natural laws is that you can start at any scale and still make a meaningful difference. Every plant you add that produces cloud condensation nuclei contributes to the atmospheric conversation. Every square foot where you slow down water and encourage infiltration helps build the foundation for larger changes.

Step #1: Assessment Phase: Begin by asking three fundamental questions about your land. First: What does your landscape currently breathe into the atmosphere? Walk around on hot days and pay attention to whether different areas feel cooler and more humid (good breathing) or hotter and drier (stressed breathing). Second: Where are your best opportunities to add more dance partners? Look for areas where you could establish vegetation that excels at producing aerosols: plants in the pine family, eucalyptus species, and many flowering plants are champions at this. Third: How can you help water slow down and stick around longer? Even simple changes like adding mulch, creating gentle berms, or establishing small rain gardens can begin transforming fast water into slow water.

Step #2: Strategy Development: Start with your highest-impact, lowest-effort opportunities. If you have any natural depressions or areas where water tends to collect, these are perfect places to establish your first 'rain garden' plantings. If you have any elevation changes, even modest ones, design your plantings to take advantage of the mountain magic law. If you have areas that are already showing good vegetation growth, expand and enhance these rather than starting from scratch in difficult locations.

Step #3: Implementation Approach: Think in terms of building systems rather than just planting individual plants. Create connected patches of vegetation that can work together, rather than scattered individual specimens. Focus on species that serve multiple functions: plants that produce lots of cloud condensation nuclei, contribute to evapotranspiration, help slow down water, and provide other benefits like food or beauty.

Step #4: Apply the Mini Magic Number: Even at small scales, you can apply the magic number principle by creating concentrated areas of high-impact vegetation rather than spreading your efforts too thinly. A quarter-acre of dense, diverse plantings will have more atmospheric effect than the same plants scattered over several acres.

Step #5: Mix Time Horizons: Remember that you are not just planting for the first year, you are designing systems that will become more effective over time. Choose fast-growing pioneer species that can quickly establish the basic conditions, combined with slower-growing climax species that will take over and enhance the system as it matures.

Starting Small, Thinking Big: You might begin with something as simple as a rain garden that captures runoff from your roof, planted with species that excel at producing aerosols and moderating local climate. Or you might start by creating a small windbreak that demonstrates how vegetation can organise and calm airflow patterns.

The key is to start somewhere and then observe how your landscape responds. Each small success teaches you more about how these laws work in your specific conditions, and each success creates foundation conditions that make the next phase easier and more effective.

Knowing When It's Working

How Do You Measure Success? The effects of working with rainfall laws show up in ways both subtle and dramatic, and learning to recognise these signs keeps you motivated and helps you refine your approach over time.

Signal #1: Feel the Air Change: The first changes you will notice are sensory. Areas where you have applied these principles will begin feeling different, softer air, more comfortable temperatures, longer-lasting humidity after rains. Take time to walk through your developing systems and pay attention to how they feel compared to unchanged areas. This is not just subjective experience, you are actually feeling measurable changes in local climate conditions.

Signal #2: Watch the Patterns, (like dew and mist): Start paying attention to how moisture behaves in and around your plantings. Do you see morning dew forming more readily? Does mist linger longer? Do you notice more bird and insect activity, which often indicates improved microclimate conditions? These are all signs that your landscape is beginning to have more productive conversations with the atmosphere.

Signal #3: Track the Cycles (like damp soil after rain): Over time, you should begin to notice more reliable precipitation patterns in your immediate area. This does not mean you will get rain when neighbouring areas do not, but you may find that when rain does come, it tends to be more gentle and productive rather than harsh and erosive. You may notice that your soil stays moist longer after rains, or that your vegetation looks more resilient during dry periods.

Signal #4: Keep a Simple Log: Keep simple records of what you observe. Photos taken from the same spots at different seasons can show remarkable changes over just a few years. Note when you first observe morning mist forming over your plantings, or when you notice the air feeling noticeably different in your enhanced areas compared to unchanged spaces.

Signal #5: Growth Wants to Expand: As you see positive results in your initial areas, you will naturally begin to see opportunities to connect and expand these systems. Success breeds success because healthy vegetation creates conditions that make it easier to establish more healthy vegetation nearby.

The most encouraging thing about this approach is that positive changes often happen faster than you might expect. While reaching the full 'magic number' threshold for regional weather influence takes time and substantial area, the local microclimate improvements that make your immediate environment more pleasant and resilient can begin showing up within a single growing season.

Remember that you are not just improving your own property, you are contributing to larger patterns that benefit entire watersheds and bioregions. Every successful application of these laws creates demonstration sites that inspire others and contribute to the broader transformation of landscapes from rain-resistant to rain-attracting systems.

Part V: The Bigger Picture

Moving from backyards to bioregions, remember that the remarkable thing about working with natural laws is that small changes create ripple effects that extend far beyond their original boundaries. When you establish a rain garden in your backyard, you are not just helping your own property, you are contributing to neighbourhood-scale changes in humidity and microclimate. When your neighbourhood develops multiple rain-attracting landscapes, you are influencing watershed-scale patterns. When watersheds transform, entire bioregions begin shifting toward more stable, resilient climate patterns.

This is not wishful thinking, it is exactly what happened in Brazil's Cerrado and what is beginning to happen in restoration projects around the world. Individual actions connect to create regional weather changes because the atmosphere does not recognise property boundaries. The laws of rainfall work at every scale, from garden to continent.

We are living through a time when understanding these connections matters more than ever. As traditional weather patterns become less reliable, the ability to participate actively in creating local climate resilience becomes not just beneficial, but essential. The vision is simple but profound: landscapes that create their own resilience rather than depending on external systems to provide what they need.

Nature's Invitation to Dance

Rain Is Not Random - It Is A Response

Let us return to that puzzle we started with: why do some places get plenty of rain while their neighbours stay thirsty? Now you know the answer is not mysterious or random. It is the predictable result of how landscapes interact with the atmosphere through laws as reliable as gravity.

The green valley that thrives while the brown hill struggles is not experiencing different weather, it is creating different weather. Through millions of dance partners floating in the air; through perfectly moderated temperatures that encourage gentle rainfall; through stable breathing space that supports organised cloud formation; through natural elevation changes that help air release its moisture; and through calm winds that allow weather systems to develop properly.

Every healthy landscape is constantly in conversation with the sky above it, teaching the atmosphere how to behave. Degraded landscapes teach chaotic lessons that result in chaotic weather. Thriving landscapes teach stable lessons that result in reliable precipitation.

This understanding changes everything. Instead of being passive victims of whatever weather patterns happen to pass overhead, we become active participants in creating the conditions for abundance. Instead of seeing drought and flood as unchangeable facts of climate, we see them as symptoms of disrupted relationships between land and sky that we can help heal.

The tools for this healing are neither complicated nor expensive. They are as simple as understanding how plants create opportunities for water vapour to condense; how vegetation moderates temperature extremes; how healthy soil helps water slow down and soak in rather than rushing away destructively. They are as accessible as observing which areas of your landscape already demonstrate good relationships with moisture and expanding from there.

Every tree planted with understanding of its atmospheric effects, every rain garden designed to capture and infiltrate water, every patch of soil restored to support robust vegetation contributes to this healing relationship between land and sky. Every small success creates conditions that make larger successes possible.

The hidden dance of rainfall has been happening around us all along, but now you know the steps. The sky is extending an invitation to participate in creating the weather patterns our landscapes need to thrive. The dance partners are waiting, the music is playing, and your landscape is ready to learn some new moves.

The question is not whether these laws work. The Brazilian Cerrado and countless other success stories have already proven that. The question is how quickly we can learn to work with them rather than against them, how rapidly we can transform our understanding from 'weather happens to us' to 'we participate in creating weather'.

This is ultimately a story about agency in the face of uncertainty, about discovering that we have far more power to influence our circumstances than we might have imagined. It is about recognising that the same natural processes that created the climate challenges we face can be harnessed to create the climate solutions we need. The dance between land and sky continues every moment of every day. Now you are invited to join in.

References

1. Bayhaqi. (n.d.). *Shu'ab al-Īmān: Hadith on withheld rain.* Hadith Collection. Retrieved from https://sunnah.com
2. Carslaw, K. S., et al. (2010). *A review of natural aerosol interactions and feedbacks on climate.* Atmospheric Chemistry and Physics. https://doi.org/10.5194/acp-10-1701-2010
3. Giannakopoulos, C., et al. (2010). *Impact of climate change on hydrological cycle and Mediterranean ecosystems.* Environmental Science & Policy. https://doi.org/10.1007/s10113-011-0219-8
4. Imam Ahmad. (n.d.). *Musnad Ahmad: Related narration on rain.* Hadith Collection. Retrieved from https://sunnah.com
5. Makarieva, A. M., & Gorshkov, V. G. (2007). *Biotic pump of atmospheric moisture as driver of the hydrological cycle.* Hydrology and Earth System Sciences. https://doi.org/10.5194/hess-11-1013-2007
6. Oliveira, R. S., et al. (2014). *Changes in rainfall patterns and the impact on Cerrado vegetation.* Biotropica. https://doi.org/10.5194/we-21-55-2021
7. Stull, R. B. (1988). *An Introduction to Boundary Layer Meteorology.* Springer. https://link.springer.com/book/10.1007/978-94-009-3027-8
8. The Qur'an. (n.d.). *Surah An-Nur (24:43).* Various translations. https://quran.com/24/43
9. Trenberth, K. E., et al. (2007). *Water and energy budgets of hurricanes: Case studies of Ivan and Katrina.* Journal of Geophysical Research. https://doi.org/10.1029/2006JD008303
10. Whiteman, C. D. (2000). *Mountain Meteorology: Fundamentals and Applications.* Oxford University Press. https://global.oup.com/academic/product/mountain-meteorology-9780195132717

CHAPTER 8

Stepwells and Ancient Practices

Vasudha Pant

"In one drop of water are found all the secrets of all the oceans; in one aspect of You are found all the aspects of existence."

~Kahlil Gibran

Vasudha Pant

As an environmentalist and agricultural scientist, conscious of a wider social context, **Dr. Vasudha Pant** has devoted her professional, personal and spiritual experience in taking refuge in, and nurturing Mother Earth. Working at reviving dying and dead rivers of Uttarakhand, her main area of operation is along the River Suyal. She has committed to investing her time and energy for the youth and ecology of Uttarakhand.

Dr. Pant is an agriculture graduate from G. B. Pant University of Agriculture and Technology. She pursued her Ph.D. at CIMAP. She has published over twenty-five research papers and has authored Beginning of Agriculture and Domestication in India, a book in the series of HIST (History of Science and Technology in India) published by US based Infinity Foundation. She has presented more than 30 papers at national and international seminars, mainly focusing on traditional knowledge systems, agriculture and sustainable development, dying water sources, history of science and technology and livelihood generation in Uttarakhand's mountain regions.

Stepwells and Ancient Practices

Vasudha Pant

Water: Civilisational Lifeline

The birth of civilisations are intertwined with the presence of rivers that have historically been the foundation for human societies to survive and thrive. Rivers have provided water for drinking; fertile land for agriculture; a means of transportation, making them indispensable for early settlements. As a result, many of the world's earliest and most influential civilisations emerged along riverbanks.

The Mesopotamian civilisation, often called the 'cradle of civilisation', developed between the Tigris and Euphrates rivers. These rivers provided fertile soil through annual flooding, enabling the growth of crops like barley and wheat. The abundance of food led to population growth, the establishment of cities like Babylon, along with advancement in trade and governance. Similarly, the Nile river played a central role in the rise of ancient Egypt. The predictable flooding of the Nile enriched the soil along its banks, allowing for the cultivation of crops that sustained the Egyptian economy and supported the construction of monumental structures like the Pyramids.

In South Asia, the Indus Valley Civilisation (c.3300–1300 BCE) thrived along the Indus river and its tributaries, driven by its geographical and environmental advantages which provided fertile soil for agriculture due to seasonal flooding. This allowed the cultivation of crops such as wheat, barley and cotton, supporting large stable populations. The surplus agricultural production facilitated urbanisation, leading to the development of sophisticated cities like Mohenjo-daro, Harappa and Dholavira.

These cities showcased advanced planning, with well-laid streets, drainage systems, granaries, and water management structures. The use of standardised weights, seals and scripts points to a highly organised socio-economic structure. The surplus agriculture production provided surplus manpower that were engaged in craftsmanship. The people got the opportunity to engage in extensive trade networks, exchanging goods like beads, textiles and pottery with Mesopotamia, Central Asia and the Persian Gulf. The economic prosperity fostered cultural and technological advancements, including metallurgy and artistic craftsmanship.

In China, the Yellow River was the lifeline of early Chinese civilisations, providing fertile soil essential for rice and millet farming, which formed the basis of their economy and culture.

Rivers were not just sources of sustenance but were also centres of cultural and spiritual significance. They shaped the myths, religions and social structures of the civilisations that arose along their banks, demonstrating the profound relationship between rivers and human progress.

The rise and collapse of civilisations has often been closely linked to the abundance and scarcity of water. Throughout history, several civilisations have faced decline when water sources became insufficient or unreliable, leading to social, economic and political turmoil.

One of the most notable examples is the collapse of the Mesopotamian civilisation. The once-flourishing cities of

Mesopotamia, which relied on the Tigris and Euphrates rivers, faced environmental degradation leading to food shortages and weakening of the economic and social foundation that eventually contributed to the downfall of this cradle of civilisation.

Similarly, the Indus Valley Civilisation, one of the most advanced early civilisations, experienced decline partially due to changes in river patterns. The drying up of the Saraswati river and the shifting course of the Indus river disrupted agriculture and trade, which were vital to the civilisation's prosperity. These environmental changes likely forced the population to migrate, leading to the abandonment of major cities like Mohenjo-daro and Harappa.

The Maya civilisation in Central America is another example of how water resources affect civilisations. Prolonged droughts, coupled with deforestation and unsustainable agricultural practices, created water shortages that strained their complex society. As water became scarce, cities were abandoned and the civilisation fragmented into smaller socio-economic groups.

These examples underscore how water scarcity, often exacerbated by human activity or climatic changes, has been a decisive factor in the collapse of civilisations throughout history. And those who do not learn from the past are doomed to repeat it.

Worst Water Crisis

Today the world faces a critical water crisis due to population growth, climate change and poor management of resources. Freshwater demand is rising while supplies are dwindling due to droughts, pollution and over-extraction. Many regions, particularly in Africa, South Asia and the Middle East are suffering severe water scarcity, leaving millions without access to clean water or sanitation. The water situation in India is very critical.

We are facing a severe water crisis, one of the most pressing challenges in the country's development trajectory. The growing demand for water, coupled with its diminishing availability and uneven distribution, has created a situation that affects millions of people across rural and urban areas. This crisis is shaped by multiple factors, including rapid population growth, agricultural dependence, climate change and poor water management practices.

India holds only 4% of the world's freshwater resources while supporting 17% of the world's population. According to NITI Aayog's Composite Water Management Index (CWMI) report, India is experiencing the worst water crisis in its history, with nearly 600 million people facing high to extreme water stress. With over 1.4 billion people, India is now the most populous country in the world.

India's annual per capita water availability was 1,486 cubic meters in 2021, which falls under the water stress category (less than 1,700 cubic meters). This is projected to reduce to 1,341 cubic meters by 2025 and 1,140 cubic meters by 2050. Rapid urbanisation and industrialisation have placed enormous stress on existing water resources.

Many cities including Delhi, Bengaluru and Chennai are struggling to meet the water demands of their expanding populations. The situation is exacerbated by outdated infrastructure and inefficiencies in water distribution, resulting in significant wastage. In rural areas, the dependency on groundwater for drinking and irrigation has led to alarming levels of overextraction. India is the largest consumer of groundwater globally. Unsustainable agriculture practices have resulted in declining water tables in many regions, especially in Punjab, Haryana and Rajasthan.

Agriculture, which consumes most of India's freshwater, is a key contributor to the crisis. Traditional practices such as flood irrigation and the cultivation of water-intensive crops strain limited resources. Simultaneously, climate change has worsened the situation by disrupting monsoon patterns and increasing the frequency of droughts and floods, further depleting water availability.

Water pollution adds another layer of complexity. Rivers such as the Ganga and Yamuna are heavily polluted with untreated industrial discharge, agricultural runoff and urban sewage. The lack of effective waste management systems and enforcement of environmental regulations continues to aggravate the issue.

Water Crisis in Uttarakhand

The Himalayan state of Uttarakhand is often referred to as the 'Land of Rivers'. Yet, it faces a paradoxical water crisis despite being the origin of some of the country's major rivers like the Ganga and Yamuna. This crisis has emerged due to a combination of natural and human-induced factors, threatening the region's ecological balance and the livelihood of its people.

Like any other region, one of the primary reasons for water scarcity in Uttarakhand is the rapid urbanisation and unplanned development that has taken place over the past few decades. To meet the growing demand, excessive groundwater extraction has become common, resulting in a rapidly declining water table. Additionally, the encroachment and destruction of natural water bodies and recharge zones has further diminished the availability of water. Traditional sources such as *naulas* and *dharas* (mountain springs), which once sustained rural and hill communities, are either drying or disappearing altogether due to neglect and deforestation.

Climate change has also emerged as a critical factor impacting the water crisis in Uttarakhand. Erratic rainfall patterns and reduced snowfall in the higher altitudes has significantly affected the recharge of rivers and natural aquifers. Glaciers, the lifelines of the Himalayan river systems, are retreating at an alarming rate due to global warming. This retreat is reducing the flow of glacier-fed rivers, impacting not only Uttarakhand but also the downstream regions dependent on these water sources. Seasonal variations in rainfall have become more pronounced, with extended dry periods leaving many areas, especially in the hilly regions, facing acute water shortages during the summer months.

Rural areas, particularly those in remote mountainous regions, are the hardest hit with women and children often having to walk long distances to fetch water for daily needs.

The burgeoning tourism industry has also played a role in aggravating the crisis. Popular destinations such as Mussoorie, Rishikesh and Nainital experience a seasonal surge in population because of tourists, putting additional pressure on already strained water resources. Hotels, resorts and other facilities catering to tourists often extract water unsustainably, leaving local communities to grapple with shortages.

Significance of Traditional Water Sources

Traditional water sources hold immense significance, both historically and currently, in their role of sustaining communities, ecosystems and cultures across the world. These sources, including springs, stepwells, tanks, ponds and reservoirs, have been an integral part of human civilisation, especially in regions where modern water infrastructure was absent. Their value extends beyond their utilitarian purpose, embodying sustainable practices, cultural heritage and ecological balance.

In the context of water management, traditional water sources were designed ingeniously to cater to the needs of local communities while aligning with the natural environment. For instance, stepwells in India, such as the famous Rani ki Vav in Gujarat, not only stored water but also helped recharge groundwater. Similarly, ancient tanks and reservoirs in Tamil Nadu, such as the Veeranam tank, provided irrigation water for agriculture while preventing surface runoff. These systems were built using local materials and knowledge, making them both cost-effective and environmentally sustainable. They ensured water availability during dry periods and helped communities adapt to seasonal changes.

Ecologically, traditional water sources play a critical role in maintaining biodiversity and regulating the local climate. Ponds and wetlands, for example, serve as habitats for various aquatic species and birds. They also act as natural carbon sinks, reducing greenhouse gas emissions. Springs in mountainous regions not only provide drinking water, but also sustain forests and agricultural fields, creating a harmonious balance between nature and human activity. The preservation of these sources ensures the continuity of vital ecological functions.

Culturally, traditional water sources are deeply intertwined with the social and spiritual fabric of communities. In many cultures, water bodies are revered as sacred and are often associated with deities or religious rituals. Traditional water sources also fostered communal harmony, as their maintenance and use required collective effort and cooperation among community members. This sense of shared responsibility reinforced social bonds and ensured equitable access to water.

Furthermore, traditional water sources highlight the importance of living in harmony with the environment. They demonstrate how human ingenuity can work in consonance with nature, rather than by exploiting it. By preserving and restoring these systems, communities can reconnect with their cultural heritage while building a more sustainable future.

The significance of traditional water sources lies not only in their ability to provide water but also in their role as symbols of sustainability, community and ecological stewardship. Recognising their value and ensuring their preservation is essential for addressing modern water challenges and fostering a deeper connection with our shared natural heritage.

Traditional Hydraulics of India

Kalyan Kumar Chakravarty writes in the chapter, *Towards a Sustainable Water Policy*:[1]

"Apart from archaeological examples, water harvesting structures continue to be maintained by indigenous communities all over the country. Groundwater continues to be drawn from aquifers or palaeo-channels in Rajasthan through *bardi* (shallow wells in depressions), t*ankas, khadin* and *johad* (underground storage pits), *kunds* (cisterns in catchments), *beris* (wells on river beds). Water is still canalised from streams or rivers into diversion, inundation or overflow channels, occasionally interspersed with storage structures and outlets, like *guhls* and *kuhls* in Kashmir, *zings* in Ladakh, *jhoras* in Darjeeling, *bandharas* in Maharashtra, *keres* in Karnataka, and *kanwas* in the flood plains of West Bengal and Bihar. In Himachal Pradesh, the *naulas* are waterscapes, centred on medieval shrines. Terracing, mulching, strip cropping, combining vegetation, cattle yards, manure dumps and ponds, continue to be in vogue in the *zabos* of Nagaland. The principle of

1. Chakravarty, Kalyan Kumar; Gyani Lal Badam and Vijaya Paranjapye, 2006. *Traditional Water Management Systems of India*, New Delhi.

gravitation and flow across slopes is still used in split bamboo drip pipes, to conduct stream and spring water in Meghalaya; to store rain water in jackwells among Shompens and Jarawas in Andaman and Nicobar Islands; to carry underground water through the system, comprising underground galleries, with hollow masonry columns, vertical air shafts, wells and tile or masonry pipes in Burhanpur, Madhya Pradesh. In the absence of perennial rivers (except Mahi in central Gujarat or Tapi in South Gujarat), Gujarat has developed 500 types of *vavs* (stepwells), *ekmukhi* to *saptamukhi* (with one to seven openings), to catch the water where it falls, to travel down to its level, to take advantage of the water conserving strata of lignite and miliolite, or of lakes, stabilised by sand dunes, and covered by vegetation. *Virdas* have been constructed in Gujarat to separate sweet from saline water. Houses have been built with tilted roofs in Rajasthan to collect and store rainwater. *Surangams* or tunnels in North Malabar collect moisture seeping through the surrounding laterite hillocks".

The Stepwells of India: Masterpiece of Ancient Engineering

India with its rich heritage and historical grandeur, is home to countless architectural wonders that stand as testimony to the ingenuity and artistry of its ancient civilisation. Among these marvels, the stepwells or *baoris, baolis,* and *vavs,* occupy a unique place. These water reservoirs, designed to harvest, conserve, and access water, are not just functional structures but also exquisite works of art that reflect the socio-religious ethos of their times. Predominantly found in arid and semi-arid regions of western India, stepwells were an ingenious response to water scarcity and are a fascinating blend of utility, spirituality and artistic expression.

The concept of stepwells dates back to as early as 600 CE, with their peak construction period occurring between the 11th and 16th centuries during the reign of the Solanki, Paramara and Mughal dynasties. While the primary function of stepwells was to store water for drinking, irrigation and religious rituals, they also served as social hubs, resting places for travellers and architectural manifestations of patronage by rulers and wealthy merchants.

Stepwells evolved in response to the climatic challenges faced by regions like Gujarat, Rajasthan and parts of Madhya Pradesh, where monsoonal rainfall was both seasonal and erratic. By creating subterranean water systems, ancient engineers harnessed groundwater, ensuring a consistent supply during dry spells. These wells were constructed with steps leading down to the water level, allowing easy access even as the water receded during the summer months.

The architectural design of stepwells reflects a harmonious blend of utility and aesthetics. Typically, they are multi-storeyed structures with a series of descending flights of steps that lead to the central water reservoir. The design ensured stability and ease of access, with the steps often flanked by intricate carvings, sculptures and pavilions.

Stepwells were more than just utilitarian structures; they were deeply intertwined with the cultural and spiritual lives of the communities they served. Many stepwells were constructed near temples and were used for ritual purification before entering the shrine. They also served as venues for religious ceremonies and social gatherings, fostering a sense of community.

Water in Indian culture is considered sacred and life-giving. The stepwells, with their intricate carvings of gods and celestial beings, were seen as sacred spaces where the divine and the earthly converged. The act of descending into a stepwell symbolised a spiritual journey, moving from the mundane world above to the tranquil depths below.

Notable Stepwells of India

India is home to thousands of stepwells, each with its own unique story and architectural style. Some of the most famous include:

1. Rani ki Vav (Patan, Gujarat)

Rani ki Vav, a UNESCO World Heritage Site, is a masterpiece of stepwell architecture built in the 11th century by Queen Udayamati in memory of her husband, King Bhima I. This stepwell is celebrated for its intricate carvings and sculptures, primarily depicting Lord Vishnu and his incarnations. With seven levels of descending stairs, it showcases unparalleled craftsmanship and an extraordinary fusion of engineering and artistry.

2. Chand Baori (Abhaneri, Rajasthan)

One of the largest and deepest stepwells in India, Chand Baori was constructed in the 9th century by King Chanda of the Nikumbha dynasty. With 3,500 narrow steps arranged in a perfect geometric pattern, this stepwell is an architectural marvel. It also features a palace at one end, highlighting its dual role as a water reservoir and a recreational space for royalty.

3. Adalaj Stepwell (Ahmedabad, Gujarat)

Built in 1499 by Queen Rudabai, Adalaj stepwell is a five-storey structure renowned for its blend of Hindu and Islamic architectural styles. The stepwell's walls are adorned with intricate carvings depicting flowers, birds and mythological scenes, reflecting the artistic sensibilities of the era.

4. Agrasen ki Baoli (Delhi)

Located in the heart of modern Delhi, Agrasen ki Baoli is an ancient stepwell believed to have been built during the

Mahabharata era and later rebuilt by the Agrawal community. With its 108 steps and three levels of arched niches, this stepwell is a striking reminder of Delhi's historical depth amidst its urban sprawl.

5. Toorji ka Jhalra (Jodhpur, Rajasthan)

Constructed in the 18th century by a queen of Maharaja Abhay Singh, Toorji ka Jhalra is a fine example of Rajputana architecture. Recently restored, it now serves as a popular tourist attraction and a testament to the community's effort to preserve their heritage.

Naulas and Dharas of Uttarakhand

The pristine mountains of Uttarakhand, often referred to as devabhoomi or land of the gods, are not only home to towering peaks and sacred shrines but also to a fascinating network of traditional water management systems known as naulas and dharas. These structures, deeply integrated with the lives of hill communities, serve as vital lifelines, supplying water in a region where springs and small streams are the primary sources of sustenance.

These water systems are not just functional but are also emblematic of the region's cultural ethos and ecological wisdom. The history of naulas and dharas dates back to ancient times, with their construction rooted in the wisdom of communities that adapted to the challenges posed by Uttarakhand's rugged terrain and limited water sources. These traditional water systems were established as a means of ensuring a perennial supply of clean drinking water, especially in regions where rivers and large water bodies were inaccessible.

Naulas and dharas are mentioned in local folklore, historical records and even inscriptions found in old temples. They are often associated with rulers, local chieftains or benevolent community leaders who funded their construction as acts of philanthropy or devotion. Over centuries, these systems have become integral to the region's way of life, forming a sustainable low-cost method of water management.

Understanding Naulas and Dharas

The Central Himalayan Region of Uttarakhand has had a plethora of rivers (glacial or spring fed), lakes, waterfalls, *gaads* (streams), and *gadheras* (rivulets) to hold or carry off the precipitation. But due to their very nature and because of their distance from habitation they could not be depended upon to sustain the needs of the local populace. Therefore, there was need to harvest water that flowed through precipitation. Naulas and dharas of Uttarakhand are an example of such traditional hydraulic technology.

The architecture of naulas and dharas reflects a deep understanding of the local geography and ecology. Their construction showcases the traditional knowledge of the communities that built them, ensuring both durability and ecological balance. The use of locally sourced stone and lime mortar not only made naulas and dharas cost-effective but also ensured their harmony with the environment. The enclosed design of naulas minimises water loss through evaporation, while dharas are often surrounded by vegetation that aids in maintaining water flow and quality.

However, despite their historical and ecological importance, naulas and dharas have faced significant challenges in recent decades. Again, the factors contributing to their decline are increasing urbanisation, migration of rural population to cities, unsustainable development practices, deforestation and climate change. The introduction of piped water supply and borewells, neglect and pollution from human and animal activities have all impacted water quality.

Naulas

The naulas of Kumaon present a grand admixture of beauty, technology and faith. Naula is one of the major hydraulic techniques that is typical of the Kumaon region of Uttarakhand. These were built by hill folks since antiquity, primarily for drinking purposes and household chores. Naulas are small, stone-lined shallow wells or water reservoirs designed to tap underground water sources, typically fed by springs or aquifers. These structures are often enclosed and covered to prevent contamination and evaporation. Naulas are located in areas where groundwater can be accessed close to the surface, making them highly effective in Uttarakhand's hilly terrain. Depending upon the size of the village, each village may have one or more naulas to meet their needs. At one time Almora town is believed to have had 360 naulas. There are naulas that are ascribed to the Katyuris, the earliest known dynasty of kings of the region.

Some examples of naulas worth mentioning are Mohana naula, Jaurasi Mandir ka naula, Shivalaya naula, Syunrakot naula, Chausli naula, Hathi naula, Katyuri naula and Hawalbag naula.

Key Features of Naulas:

Construction: Entire structure is below the surface of the ground. Traditionally at the bottom of the pit a copper plate is fixed. The lowest step is usually 60x60 cm. Naulas are built using locally available stones. It is interesting to note that the number of steps in a naula is invariably an odd number such as 5, 7, 9, 11 and so on. Generally the topmost step comprises an area of about

150x150 cm. The naulas are often adorned with intricate carvings or inscriptions. They are usually square with a stone-lined basin that collects water.

Design: The structures are designed to ensure purity and sustainability. A small opening allows access to the water, while the rest of the structure is enclosed to prevent pollutants like dust and leaves from entering. The steps of naulas are not meant to be used to enter inside.

Religious Significance: Naulas are considered sacred and are built inside a temple-like structure. The place is often regarded as holy and used for rituals also.

Dharas

Dharas, on the other hand, are natural or artificially channelled surface water that flow continuously from springs or melted snow. These are typically open and accessible, providing water for drinking, agriculture, and livestock.

Key Features of Dharas:

Location: Dharas are found on slopes where natural springs emerge. They are often surrounded by vegetation, which helps maintain water quality and quantity.

Community Utility: Dharas serve as communal spaces where villagers gather water and engage in social interaction. They are also used for irrigating small fields and providing water for animals.

Sustainability: Like naulas, dharas are closely linked to local ecosystems. Their maintenance often involves the preservation of nearby forests and catchment areas.

The ecological role of naulas and dharas is crucial in recharging groundwater by collecting and storing rainwater. These water systems are often located near forests, creating interdependence between the water source and the surrounding vegetation. Villagers traditionally protect these forests to ensure the sustainability of the water supply. The presence of naulas and dharas helps maintain a cooler and more humid microclimate, which is beneficial for local flora and fauna.

In an era of increasing water scarcity and climate change, the stepwells of Gujarat, Rajasthan or the naulas and dharas of Uttarakhand, offer valuable lessons in sustainable water management. Traditional hydraulics are not merely water sources; they are symbols of the region's harmony with nature, cultural identity and ancestral wisdom. They stand as enduring symbols of human ingenuity, resilience and artistry. They are a testament to the harmonious relationship between nature and architecture, practical utility and spirituality. Their ability to recharge groundwater and provide year-round water access serves as a reminder of the importance of traditional wisdom in addressing modern challenges. Reviving and adapting them could play a crucial role in contemporary urban planning, particularly in water-stressed regions.

While they face threats from modern development and environmental degradation, their revival holds immense potential for sustainable water management and cultural preservation. By honouring these traditional systems, humanity can ensure a future where people remain connected to their heritage while securing a critical resource for generations to come.

CHAPTER 9

Adapting Agriculture to Diminishing Water

Efforts in the Himalayan Region

Suman Sahai

"The earth, the air, the land and the water are not an inheritance from our forefathers but on loan from our children. So we have to handover to them at least as it was handed over to us."

~Mahatma Gandhi

Suman Sahai

Dr. Suman Sahai, who has had a distinguished scientific career in Genetics, is a recipient of the Padma Shri, the Borlaug Award, Outstanding Woman Achiever awards, the Birbal Sahni Gold Medal and the Order of the Golden Ark of the Netherlands.

She is founder Chairperson of Gene Campaign, a leading research and advocacy organisation, working on food, nutrition and rural livelihoods. Dr. Sahai, has published extensively on science and policy issues, including 13 books, 17 book chapters and several hundred articles.

Dr Sahai chaired the Planning Commission Task Force on 'Agro biodiversity and Genetically Engineered Organisms', for the Eleventh Plan. She has served on the Research Advisory Committees of national scientific institutions and has been a member of several national policy forums on scientific research and education, biodiversity and environment, biotechnology and bioethics as well as intellectual property rights.

Adapting Agriculture to Diminishing Water
Efforts in the Himalayan Region

Suman Sahai

Some years ago, I was watching a James Bond film called *Quantum of Solace,* when it struck me that water, as a sought-after resource, had caught the attention of the popular imagination. Unlike earlier Bond villains who dealt in such lucrative commodities as drugs, gold, diamonds or atomic bombs and nuclear missiles, in this 2008 film, Dominic Greene, the main villain plots to monopolise the entire water supply of Bolivia in order to establish control over the country. A Bond film with water as its central theme was certainly an unusual way to draw attention to the growing water scarcity at the global level, and a portend of the coming water wars.

Now as climate change kicks in and availability and access to water diminishes further, water being a subject of wars becomes a depressing reality.

In our own country we have a long tradition of conserving water because it was not always plentiful. Ancient India was replete with water tanks and lakes; check dams and anicuts; sophisticated aqueducts bringing water from far away; and elaborately constructed underground baolies.

Modern India has lost these wise conservation policies and instead adopted an expansive utilisation of water, increasing supply as demand increased, without paying any attention to managing demand. Today, we have contestation over water resources everywhere, for instance the Cauvery dispute between Tamil Nadu and Karnataka or the Teesta water sharing dispute between India and Bangladesh. Clearly, we need to go back to policies that husband water and use it sustainably. This will have to happen at all levels that water exists and all the colours it is coded by.

The Colour of Water

Curious as it may sound, water is categorised in different colours: *blue*, *green* and *grey*. Blue water is the one we are most familiar with and use most consciously. This is the water that is found in rivers, dams, lakes and ponds. It is also the groundwater which exists in aquifers and the water we see in marshes and wetlands. Blue water comes from rain and snow and also when glaciers melt and feed rivers. It is the water we use for drinking, for domestic and industrial use and also for irrigation and food production. In fact, at present the biggest portion of blue water (70%) goes to agriculture and food production.

Green water is the invisible water to which we need to pay more attention. It is the water found in soil, that is used by plant roots to distribute through capillary action, the very basis of plant growth. It is also used by the web of organisms and microorganisms that live in the soil that are responsible for keeping the soil 'living' and fertile.

Green water comes from rainfall that is soaked by the soil. For a healthy agriculture system, it is essential to keep green water in the soil and not lose it through evaporation. Ways to keep the moisture content in the soil high is to prevent over tilling the soil which exposes it to the sun repeatedly. Other methods are mulching and bunding of fields and crop production areas. Mulching can be done by growing cover crops, covering by plastic sheets or by spreading a layer of straw or leaves over the soil to prevent evaporation. Contour bunding on the other hand is a proven sustainable land management practice for sloping land and terraced fields in mountain regions where the soil has poor water retention capacity and productivity is low. Bunding is done by making small soil embankments along the field edge to intercept water runoff flowing down.

According to Falkenmark and Rockstroem, countries that have less than 1300 cubic metres per capita of combined blue and green water, would be unable to produce a balanced diet of 3000 kcal per capita per day, a view endorsed by the Food and Agriculture Organisation (FAO).

The remaining water category is grey water or waste water. This is essentially blue water that has already been through domestic or industrial use.

Grey water is currently wasted for the most part, but should be recycled and re-used. The volume of grey water coming out of homes, offices, hospitals and industrial establishments is enormous. Most of it is discharged untreated into our blue waterways, contaminating them and causing serious harm to aquatic biodiversity. Treating this water and reclaiming it for agriculture will help to offset the anticipated shortage of water as rainfall and snowfall diminishes. Recycling and reusing grey water becomes crucial in light of the impending water crisis that the world is staring at, thanks to climate change.

Groundwater

We have been brutally wasteful with groundwater with gushing tube wells running for hours, irrespective of requirement. As a result of this waste, many aquifers have been drained dry. Over decades, Punjab has grown paddy using groundwater. Now its aquifers are mostly classified as 'dark' or empty.

Great damage was done by the politically motivated and environmentally disastrous decision to provide free electricity to farmers. The electricity supply was erratic and because no one knew when this free electricity would be available, farmers kept the tube wells switched on so as not to miss the window when the electricity was there. This s ingle flawed policy has resulted in groundwater being depleted to the point of exhaustion.

This need not have been the case. In a *jan sunwai* that Gene Campaign conducted in 2007 together with Madhya Pradesh Kisan Sangharsh Samiti (MKSS) and INSAF, farmers explicitly stated that they were happy to pay for electricity to operate their tube wells but to please allot a time when the electricity would be available.

Long after the damage has been done, many Indian states have finally enacted a groundwater law. A total of twenty-one states and Union Territories have adopted and implemented a law to regulate the use of groundwater. These are Andhra Pradesh, Assam, Bihar, Goa, Haryana, Himachal Pradesh, Karnataka, Kerala, Maharashtra, Nagaland, Odisha, Punjab, Telangana, Uttar Pradesh, West Bengal, Chandigarh, Dadra and Nagar Haveli and Daman and Diu, Jammu and Kashmir, Ladakh, Lakshadweep and Puducherry.

Climate Change and Water

Loaded on the already growing water scarcity, climate change has emerged as a magnifier of existing water problems. In the last years, with the accelerated rate at which climate change has begun to manifest itself, we have become acutely aware of the sharp decline in the availability of water.

The Indian monsoon has reduced in total volume and become more erratic. The south west monsoon which extended from June to September and the North East monsoon from May to February, have both shifted their timings in unpredictable ways that are impacting the established rhythm and practice of agriculture in both northern and peninsular India.

Because of reduced water availability, tropical areas are slated to see a contraction of 31-51 million ha of favourable cultivation areas and a significant reduction in food production in the most vulnerable areas where population density is high and food is already scarce. This will result in worsening hunger and poverty if adaptive measures are not adopted immediately.

Adaptive measures to ensure continuing food production cannot be 'one size fits all'. These will have to be tailored for each ecosystem, to suit their peculiar requirements. India is home to several diverse ecosystems, rising to the Himalayan mountain ranges from the vast plains of the Indo Gangetic basin, the biodiverse Western Ghats, the plateau region of peninsular India, and the coastal areas along India's long coastline. Climate change will have different kinds of impacts in these diverse regions. For instance, large parts of Rajasthan, Andhra Pradesh, Gujarat and even some parts of Odisha and Uttar Pradesh face frequent drought, whereas some 40 million hectares of land in the north and north-eastern belt are prone to recurrent floods. All this variability will have to be catered for.

Water in the Himalayan Region

Significant disturbances are already being seen in the Himalayan region as a result of climate change. According to ICIMOD

(International Centre for Integrated Mountain Development) by 2050, temperatures across the mountains of the Hindu Kush Himalayan region will increase by about 1-2°C, going up to 4-5°C in some places.

Glaciers are melting at unprecedented rates and rising temperatures are leading to earlier and faster spring snow melt. The Indus, a snow fed river is facing the brunt of these changes and to a somewhat lesser extent so is the Ganga. The melting of glaciers changes availability of water in Himalayan rivers and affects their flow. This is beginning to impact the behaviour of the monsoon and affect downstream regions.

The monsoon is expected to become longer and more erratic, disrupting the cycle of water availability. The trends vary, for some areas fewer extreme rainfall events are predicted with these becoming more intense and destructive, for other areas more frequent extreme events are projected. Traditional high monsoon regions like North and Northeast India, the Indo-Gangetic plains, and the Indian Himalayan region have been experiencing a decrease in monsoon rains for the past ten to fifteen years.

Changing depths of mountain glaciers and seasonal melting patterns are already having an enormous impact on water resources with the expectation that this will worsen as temperatures rise further. The increased glacial melt in the Himalayas has created vast unstable lakes. If these burst, as they are prone to do, inundation of towns and villages downstream can cause havoc. On last count, remote sensing showed 44 glacial lakes in Nepal and Bhutan.

As glacial melt increases, river flows are set to increase in the short term. There will however, be higher variability in the flows and greater volumes in the pre-monsoon months. This will lead to more frequent and unpredictable floods and droughts, negatively affecting agriculture and livelihood security in riparian regions.

Another outcome of higher temperatures in mountain areas is that instead of winter snow, there is more winter rain. This has two negative fallouts, particularly for forests and orchards. Snow sits longer on the ground, melting slowly and nourishing the roots and recharging aquifers, adding to the volume of essential green water in the soil. Rain on the other hand flows away faster causing greater soil erosion, failing to trickle into aquifers or recharging springs. Intense bursts of heavy rains carry away more sediment and soil nutrients than snow melt, all of this makes agriculture less productive.

Impact on Agriculture

Crop production systems have developed corresponding to the rainfall patterns of the region. When this pattern is disrupted, as it is now and there is a greater variability in the intra rain period, there is a negative impact on crop production.

Take for example the millet crop in Uttarakhand. Here nurseries of finger and barnyard millet are traditionally planted end May to beginning June. The monsoon rain started in the first week of June and nurtured the nurseries. The seedlings were transplanted late June. Now the monsoon comes by end June to early July and the whole millet cultivation cycle is disrupted.

Precipitation patterns are changing across all ecosystems with unseasonal rains leading to unpredictable floods and droughts. The southwest monsoon, India's lifeline for food production has diminished by 15%. For a country where food production depends largely on the monsoon, 15% is a substantial reduction. At the same time rainfall intensity is rising, increasing surface runoff which carries away the nutrient rich top soil.

Gene Campaign's Work

Recognising that one of the most debilitating impacts of climate change in the Himalayan region is the reduction in the already constrained availability of water, Gene Campaign (GC) began work on making agriculture more resilient under the changed conditions. This work has been in villages of the Kumaon region of Uttarakhand.

Survey on climate change

Around 2010-12, the organisation began to discuss Jalvayu Parivartan (Climate Change) with the community alongside our usual meetings and training programs. People were noting the erratic and weaker monsoon and the warming winters but were not sure what to make of it. In 2012, GC conducted a structured survey on people's perception and response to climate change. Not surprisingly, the diminishing availability of water was flagged as a significant area of concern affecting both domestic and agricultural use. Our work to build resilience into agriculture plus food and nutrition availability in a water deficit future has been on a number of axes.

Rainwater conservation

GC established Jal Kunds or rainwater tanks in participatory mode with rural families. These tanks, lined with high density polyethylene sheets, store approximately ten to twelve thousand litres of water. Carefully used, they last long and any tears can be easily repaired. Water from these *jal kunds* supports vegetable cultivation and can provide life-saving irrigation in a drought like situation. When required, they also provide drinking water for family cattle. GC has till date helped set up rainwater tanks which have conserved around 3,43,00000 litres of water.

Reviving crops that can provide food under water deficit conditions

Millets that were traditionally staple foods in the Himalayan region have been largely displaced in India by the subsidised wheat and rice provided through the Public Distribution System (PDS). To revive millets as a primary source of food in the future, GC began with 'awakening' programs about the value of millets under water deficit conditions, focusing on the knowledge that communities already have. The strategy to revive millets focused on finger millet and barnyard millet and was on several fronts.

Introducing greater genetic diversity in the field

By making available a wide range of finger and barnyard millet varieties to farmers, both traditional and those bred in research institutions, genetic diversity was introduced. Greater genetic diversity is a risk insurance against crop collapse under turbulent weather conditions because it is based not on one variety but a few different ones. Whatever the conditions, some variety is likely to succeed.

- To further increase genetic diversity, GC tried to reopen old channels of seed sharing. A *Beej Mela* was organised inviting farmers with their seeds from different villages. It was heartening to see the practice of seed exchange starting again with farmers exchanging notes and seeds.
- Introducing new agronomic practices to increase yield using less water. To achieve this, GC adapted the techniques used in SRI (System of Rice Intensification), chiefly spaced planting of older seedlings, to millets. Like in rice, this approach showed higher yield in many millet varieties.
- Farmer Field Schools were conducted to provide training in good agronomic practices

- Value addition to make cultivation more attractive. As an incentive to take up millet cultivation after a period of neglect, GC provided training in value addition of millet products. Marketable products like snacks and malted millet drinks give value-addition to the crop.

Increasing the Food Basket

Possible slack in food production due to extreme climate conditions and critical water deficit need to be compensated. A drop in food production is one of the fallouts of erratic rainfall. GC has established a series of homestead gardens growing vegetables. These are secured by the water from *jalkunds* and provide additional nutritious food to the family. The 1054 vegetable gardens established have produced nearly 300 tonnes of vegetables so far.

Increasing the water retention capacity of the soil

Adding biomass

Soils in the higher mountain areas are lacking in humus and biomass so are unable to hold the essential green water necessary for sustained plant growth. To improve the water retention capacity of the soil, GC has provided training in composting and established vermicompost units that have produced some 61 lakh, 46 thousand kg of compost so far.

Bunding on field borders

GC did several demonstrations on bunding in farmers' fields showing how to retain rainwater in the field and prevent the fertile top soil being washed away. Unfortunately, this technique has not been well accepted and farmers by and large do not bund their fields.

Mulching

Mulching is another technique to retain soil moisture and also controls the emergence of weeds. GC conducted demonstrations in farmers' vegetable plots spreading fallen leaves over the soil. By checking evapotranspiration, the soil is able to hold its green water for a longer period.

Clean drinking water

One of the fallouts of reduced river flow due to diminishing feed from glacial melt, is anticipated to be more turbid water with a higher level of contamination. GC demonstrated a simple, proven technique to purify such water and make it fit for drinking by using a clean cotton sari folded four to five times and filtering the water through that.

India's New Water Policy

In response to demands from many quarters that the old water policy was unable to meet the modern challenges to water use, India has finally got a new water policy, developed after stakeholder consultations and the involvement of water scholars like Mihir Shah and other experts. This 'new water governance paradigm' described in the new national water policy is based on a multi-disciplinary and multi-stakeholder approach to water management.

The New Water Policy can be considered a work in progress and can only improve as knowledgeable people continue to contribute recommendations. An important input comes from Bandyopadhyay who makes the point that Himalayan rivers must be the lynchpin of India's New Water Policy. The Indian Himalayan Region (IHR) is the water tower of Asia from where the Ganga, Brahmaputra, Indus and the Barak emanate. In addition, these high mountain ranges act as a barrier to the flow of moisture bearing winds, thus influencing the rainfall in Asia and hence the water capacity of the Himalayan rivers. In return, this has a direct bearing on the water supply and food security of northern India. Conservation and sustainable use of our water resources is the only way of ensuring an adequate food supply going ahead.

References

1. Falkenmark M, J Rockstroem 2013, Balancing Water for Humans and Nature: The New Approach in Ecohydrology – https://www.researchgate.net/publication/285831080_Balancing_water_for_humans_and_nature_The_new_approach_in_ecohydrology
2. FAO, The Contribution of Blue Water and Green Water to the Multifunctional Character of Agriculture and Land. Background Paper 6: Water – https://www.fao.org/4/x2775e/x2775e08.htm
3 Jan Sunwai on The Present Agrarian Crisis: A Report – https://genecampaign.org/wp-content/uploads/2014/07/THE_AGRARIAN_CRISIS_IN_INDIA.pdf
4. Sahai S, 2022 – The Crisis of Climate Change: Weather Report, ed. Ravi Agarwal and Omita Goyal. First South Asia Edition 2022, Routledge, Taylor & Francis.
5. Adaptation and Resilience Building. ICIMOD – http://www.icimod.org/hicap/?q=4779
6. Mool PK, Wangda D et al 2001, Inventory of glaciers, glacial lakes and glacial lake outburst floods: monitoring and early warning systems in the Hindu Kush-Himalayan region, Bhutan. 2001, ICIMOD, UNEP
7. Sahai S, 2012. Impact of Climate Change on Agriculture and Food Security. Presentation in National Seminar on Food Security: Issues, Challenges and Policy Measures, Ludhiana, 10 October. https://genecampaign.org/
8 New Water Policy, https://www.pib.gov.in/Pressreleaseshare.aspx?PRID=1782283
9. What India's New Water Policy Seeks To Delivery. Mihir Shah, 2021. https://indianexpress.com/article/opinion/columns/what-indias-new-water-policy-seeks-to-deliver-7595819/
10. Bandyopadhyay J., 2020. https://dialogue.earth/en/water/himalayan-rivers-must-be-lynchpin-of-indias-new-water-policy/

CHAPTER 10

Industry or Citizens: Who Needs Water More?

Anjan Ray

"Thousands have lived without love, not one without water."

~W.H. Auden

Anjan Ray

Anjan Ray, a scientist who trained under the guidance of Nobel Laureate Alan MacDiarmid, has worked for over 35 years in industry, academia and government across a wide range of functions. He is widely published and cited in academic research articles and book chapters and is the co-inventor for more than 10 international patents. Over 30 technologies and products have been commercialised by him and his teams during his career.

His work in sustainable energy led him to the Indian Federation of Green Energy Jury Choice Award for Outstanding Green Energy Activist in 2021. In 2023, he was conferred the IIT Delhi Alumni Award for Outstanding Contribution to National Development.

Dr Ray has also been actively involved in print and broadcast media, consumer rights, cultural education, and environmental conservation for over four decades.

Industry or Citizens: Who Needs Water More?

Anjan Ray

Some enduring memories of my early childhood involve water bodies. I was born the year after Rachel Carson shook global consciousness with her seminal book, *Silent Spring,* about the potential devastation that could be caused by a single industrial chemical that contaminated water, soil and all life forms that inhabited these.

My father served in the Navy and spent a significant part of his military career aboard ships as a medical officer. Itinerant as military life was, with postings to different stations across India every few years, one constant was his family home at Halishahar in North 24 Parganas district of West Bengal. He had four brothers and six sisters; the place was always bustling with some of us thirty-three cousins visiting for one reason or another. A pond teeming with fishes and lined with rich bamboo groves within our compound was the main source of water for the house. The Hooghly, a distributary in the Ganges delta, was barely a hundred meters away on the west, across the main road. I do not recall ever drinking any water other than from the pond or the river back then, though alum used to be added to the large water cisterns behind the house for settling suspended solids. At night, the sky was resplendent with stars; our elders taught us the shapes and names of the constellations, albeit in Bengali.

Fast forward sixty years. The house itself is gone; family members moved away to different cities in India and abroad. The pond was filled in to enable more construction; a featureless, nondescript block of apartments has taken the place of the stately old house. The river is dark green and turbid. Pigs wander around freely at its edge, nosing for edible titbits at the crematorium on the banks. About 2 km across the water, the 335 MW coal-fired Bandel thermal power plant belches its exhaust into the air. The Air Quality Index is a bit over 150; one can barely make out Venus and Jupiter at night.

20 km away in Haringhata, close to the campus of the Indian Institute of Scientific Education and Research (IISER), sexagenarian locals talk about the time that deer roamed free in the area and occasional leopards terrorised villages at night. Development, as we know it, is everywhere. Highways and interior roads criss-cross this part of the North 24 Parganas district; we relied on our own two feet, rickety bullock carts and aged careening two-wheelers back then. But all agree that one thing has changed – water quality is worse, and everyone has installed water filters. Many of them say with pride that they have the latest Reverse Osmosis (RO) unit. What sense does that make, I wonder, in an area that was once replete with freshwater and should have required no removal of dissolved solids and salinity? Is it the high levels of arsenic and fluoride reported in this district? Or is it the relentless marketing by RO manufacturers, persuading consumers to spend substantially from their pockets for clean drinking water?

In 2010, the United Nations General Assembly recognised clean drinking water and sanitation as a human right in Resolution 64/292. The resolution calls on countries to provide safe, affordable, and accessible drinking water and sanitation for all. Article 11(1) of the International Covenant on Economic, Social and Cultural Rights states that access to safe drinking water and sanitation is a human right. The 6th Sustainable Development Goal (SDG) is to ensure the availability and sustainable management of clean water and sanitation for all. In the Indian Constitution, the Right to Life is enshrined in Article 21; while a healthy person might be able to survive for three weeks without food; without water, that window shrinks to just three days. Contaminated water does not just induce diseases – including cancer – but also compromises immunity and increases morbidity and mortality, especially in children.

A particularly insightful study, published in 2021 by Paul and Das of West Bengal State University based in the same district, captured three primary sources of water contamination in North 24 Parganas, which has the highest population density and industrial concentration after the state capital, Kolkata, in West Bengal:

a. indiscriminate use of inorganic fertilizer and synthetic insecticides
b. rampant uncontrolled discharges of untreated municipal solid and liquid wastes
c. inherent susceptible character of the region due to its deltaic environment.

Between household discharge and industrial effluent into rivers and groundwater, the difference is stark. While approximately 44% of wastewater generated by households worldwide are not treated, 80% of all industrial and domestic wastewater on a global level is released into water resources without prior treatment. This clearly shows that industries are a major concern area regarding safe drinking water for citizens. Especially in the Global South, governments are not being held adequately accountable, either financially or legally. On the other hand, even municipal sewage treatment is grossly inadequate in population-dense countries like India as solid waste can leach into groundwater or contaminate rivers directly. A recent study showed that municipal solid waste from India contributed to 10% of gross impact into the world's rivers in 2020.

In general, industries across India, as well as most places around the world, tend to gravitate towards coastal riverine areas for access to water resources and shipping channels. Coastal industries are increasingly being regulated to desalinate seawater for large-scale use. While this reduces the burden on deltaic and estuarine freshwater resources and on groundwater extraction, widespread desalination is slowly pushing up local salinity as concentrated brine rejects from desalination find their way back into coastal waters, affecting marine biodiversity in favour of salt-tolerant organisms at the bottom of the food chain. Industrial impact is not the only contributor; seawater ingress is also a factor. It is estimated that India has about 1.3M ha of saline soil within arable lands in all coastal districts, with Gujarat, West Bengal and Andhra Pradesh - the top three-salinity affected coastal states – accounting for almost 90 percent of this affected area. This not only affects crop yields and agricultural income; it also directly impacts the availability and quality of drinking water for residents in these zones.

Inland industries, on the other hand, depend entirely on surface water (rivers, lakes and ponds) and draw on groundwater as these sources are in short supply, either seasonally or perennially. At the same time, many industries – especially in poorly regulated zones, often operating illegally in residential areas – do not treat their liquid effluent at all. These effluent streams find their way into sewers and gutters and eventually to streams, tributaries and major rivers – or into groundwater.

Thus, surface water and groundwater depletion has proceeded at a rapid pace in line with industrialisation of the Indian economy. Effluent-driven contamination of water used by citizens at large also continues to rise. With a shrinking denominator of total available water and a rising numerator of contaminated water resources, we have a double whammy to deal with.

This is especially true of citizens who live in catchment areas of industrial wastewater discharge and groundwater extraction. An April 2024 report speaks of villagers in the National Capital Region in Northern India migrating towards urban settlements in search of cleaner and safer drinking water, the groundwater aquifers in their rural homes having been contaminated by untreated factory discharges. Not that migration to urban centres necessarily solves water issues. On April 30, 2023, a gas leak from a manhole of a sewer line killed 11 people in Ludhiana. An investigation report issued about 9 weeks later did not find any individual or organisation directly responsible. However, as high levels of hydrogen sulphide were detected after the incident, the possibility of undetected industrial acid discharge into a sewer filled with untreated sulphur-containing organic matter could not be ruled out.

One tends to think of large, smoke-spewing factories as representative of 'industry' – but the truth is far from this perception. All manner of industries – small, medium and large – across multiple sectors, contribute to withdrawal of water from surface and groundwater sources. Given India's status as a rapidly developing economy with the world's largest population, it is not surprising that the basics – food (agriculture, agro-processing and downstream products), clothing (textiles), shelter (construction) and energy, contribute significantly to water usage as well as water contamination in the nation's quest for economic growth.

Agriculture is the most water-intensive 'industry' in India, accounting for over 80% of the country's freshwater consumption. More than half of the water used in agriculture in India comes from underground sources, which underpins depletion of groundwater levels. Rice, sugarcane and cotton are among the most water-intensive crops; while rice and sugarcane are primarily food crops, sugarcane is progressively a contributor to the ethanol industry value chain for biofuel, chemicals and liquor. A 2017 report indicated that water usage per ton of crop was 2-4 times higher in India than in peer countries.

The textile industry cumulatively across the country produces a large volume of effluent that contains a complex and varying cocktail of chemicals, dyes and other pollutants. In the water-scarce Rajasthan desert, the waters of the region's most significant river – the Luni (variously called the Maru Ganga) and its tributary, the Bandi – have been rendered virtually unfit for human consumption by the accumulated untreated discharges of the textile processing units in the Pali and Jodhpur districts. Compounding the problem, the discharges tend to render local soil infertile, affecting the yield, nutritional value and toxicological safety of crops grown in the area.

Thermal power plants, critical for India's energy infrastructure, operate on the principle of a fuel, coal or natural gas, being fired to generate steam, which then turns a turbine to produce power. Almost 70% of India's industrial water consumption comes from this sector.

Iron and steel, non-ferrous metals, petroleum refineries, pulp and paper – almost every sector that plugs into the value chain of consumption by citizens is water-intensive.

The beverage industry requires water for the liquids themselves, as well as for the farmed products used in additives, such as barley, sugar, coffee, and fruit. The bottled water sector, a lifeline for tourists and affluent citizens who rightfully consider tap water as a health risk, typically uses reverse osmosis (RO) as part of the technology suite for purification. A typical industrial RO plant wastes ('rejects') at least two-thirds of the water that it takes in, and this wastewater – concentrated in impurities – may find its way back to our streams or groundwater aquifers.

India has a suite of well-intended and carefully drafted laws with regard to both use of water and discharge of effluent by industries. The Water Prevention and Control of Pollution Act, 1974, governs wastewater treatment and pollution control. The Environment (Protection) Act, 1986, provides overarching legal authority to monitor, regulate, and penalise industrial pollution. The Factories Act, 1948 (Amendments), mandates safety and environmental standards for factory operations.

The core issue, however, is that of enforcement. A multiplicity of government agencies at both central and state levels are responsible for monitoring and enforcing regulations regarding the consumption of water and the treatment of wastewater discharges from factories. This means that neither a citizen nor an industrialist can easily identify or access the right official for guidance in matters of multi-stakeholder long-term water woes – and nor are they incentivised or supported by a complex and ambiguous system:

A. Central Pollution Control Board (CPCB), which operates under the Ministry of Environment, Forest, and Climate Change (MoEFCC), develops guidelines, standards, and policies for the treatment of wastewater and prevention of water pollution. It also monitors compliance with the Water Prevention and Control of Pollution Act, 1974, conducts inspections and assessment of industrial effluents and is supposed to ensure adherence to prescribed effluent standards.

B. State Pollution Control Boards (SPCBs), the state-level counterparts of the CPCB, are tasked with implementing and enforcing pollution control laws within their jurisdictions. They grant Consent to Establish (CTE) and Consent to Operate (CTO) for industries, specifying conditions for wastewater treatment and discharge at each industrial unit. Regular inspection of industrial facilities, analysis of effluent samples, and imposition of penalties for non-compliance fall within their ambit.

C. National Green Tribunal (NGT), a specialised judicial body to handle environmental disputes, including cases related to industrial wastewater pollution, has the authority to direct corrective actions, levy fines, and order compensation for environmental damage caused by untreated discharges.

D. Ministry of Jal Shakti focuses on the conservation of water resources, including ensuring the quality of surface and groundwater. This Ministry oversees projects and policies to prevent water pollution and promote sustainable water management.

E. Central Ground Water Authority (CGWA) regulates the extraction and use of groundwater by industries, ensuring that untreated wastewater does not contaminate groundwater resources. It is mandated to enforce guidelines for industries to treat and reuse wastewater to minimise water wastage.

F. Municipal Corporations and Local Bodies oversee urban and industrial wastewater treatment plants within their jurisdictions. They are expected to enforce compliance with local bylaws and environmental regulations regarding wastewater discharge into municipal drains or water bodies.

G. Coastal Zone Management Authorities (CZMAs) monitor industries located in coastal regions to ensure compliance with wastewater treatment and discharge norms under Coastal Regulation Zone (CRZ) notifications.

These agencies are expected to 'collectively' ensure that wastewater from factories is properly treated before discharge, thereby protecting the environment and public health. However, accountability for, and effectiveness of the mandates, as well as actions by these agencies is grossly inadequate. Bandwidth at the ground level is insufficient to police the sheer numbers of MSMEs and massive sizes of many large-scale industrial units. The complexity of the regulatory framework requires several agencies to work together, but mechanisms for such coordination vary widely across states. Opportunities for collusion in compliance avoidance between operating industries and monitoring officials are rife.

Apart from the evident lack of adequate resources for monitoring and enforcement, there is significant resistance from industries – especially smaller ones – with regard to compliance due to high costs of wastewater treatment infrastructure. The first Common Effluent Treatment Plant (CETP) for an industrial cluster, intended to help multiple MSMEs share the burden of compliance for water discharge management, was set up in Andhra Pradesh in 1989. Widely adopted across many states, this model was again well-intended, but assumed blithely that the

CETP operator would ensure discharges met norms. Evidence to the contrary is seen in a recent study from Ahmedabad, the capital of Gujarat, one of India's most industrialised states. The coefficient of reliability in terms of compliance of effluent parameters such as chemical oxygen demand (COD), biochemical oxygen demand (BOD), ammoniacal nitrogen, and phenolic compounds was shown to widely across seven CETPs that discharged their processed effluent into the Sabarmati river, clearly indicating improper functioning in a majority of these CETPs. Over-reliance on self-reported compliance data from industries and CETPs distorts the disconnect between reported figures and ground realities. Further, 'green-washing' in terms of sustainability reporting remains a real risk. For instance, a renowned multinational in South India was allegedly implicated in groundwater depletion, attracting international media attention. Another large corporation faced public ire for water and soil pollution and, among other lapses, failed to provide groundwater reports.

Ironically, the rain-fed Sabarmati river had virtually run dry in the late twentieth century, and was rejuvenated in 2002 through water supply from the larger Narmada river to its south through a canal connected to the Sardar Sarovar Dam. This means that any fluctuations in the water volume or quality of the Narmada, which has seen its own share of water contamination challenges over decades due to rapid development along its upper reaches and rapid industrialisation, now transfers into the Sabarmati.

Citizens' need for water is directly tied to their health and well-being. Water is necessary for daily survival, preventing dehydration, and maintaining hygiene, which helps control diseases. In developing regions, lack of clean water can result in severe public health crises, including outbreaks of cholera and other waterborne illnesses. When citizens do not have access to sufficient water, it affects their ability to lead dignified lives and contributes to poverty and inequality. Thus, ensuring adequate water for domestic use must be prioritised over industrial consumption. Let us not forget that it is less than a quarter of a century since Mahesh Gupta launched the country's first reverse osmosis water purifier because his children contracted jaundice. While growing up in the 1960s and '70s I do not recall drinking water from anything other than the home tap or – during dry spells – from a storage vessel with occasional alum (fitkari) treatment to settle sediments. The urban residents of 21st century India do not move into a new home without a water filter being installed first.

India's capital city (and National Capital Region) is a classic case in point. The Yamuna river, considered sacred since time immemorial, transits through Delhi for about 2 percent of its total length of 1376 km from its source at Yamunotri in the Himalayas to its confluence with the Ganga at Prayag Sangam. This section alone contributes close to 80 percent of the pollution load; the river enters Delhi at moderate water quality, which declines rapidly in this short stretch as untreated sewage discharges and impaired flow due to silting, erratic opening of barrage gates, aquatic plant growth and rampant debris disposal from unbridled construction in the area. At the Kalindi Kunj bridge in South-East Delhi, the visual and olfactory condition of the river is appalling. It is dark and murky, and a seemingly toxic foam persists for days during the dry season when the flow of the once-majestic river is virtually stagnant. Regular agricultural activities on the floodplains, shantytowns on its banks and occasional mega-events that leave behind anthropogenic wastes and non-biodegradable debris like plastics and aluminium foil, contribute to the dismal state of affairs. Roughly 2 billion US dollars (an estimated Rs 8000 cr at USD/INR exchange rates ranging from 30 in 1992 to 80 in 2024) has been spent in various initiatives to ameliorate the situation, but apparently with little positive outcome.

Among the flagship projects funded by Government agencies is the clean-up of the Barapullah drain, a natural stream once considered a tributary of the Yamuna, over which early 17th-century rulers built an iconic bridge. The drain, roughly 100m wide and 16km long, collects domestic sewage and industrial effluents and discharges an estimated 1.25 million litres of this wastewater daily into the river. Concentrations of heavy metals (lead, zinc, copper), biochemical oxygen demand (BOD), and dissolved oxygen (DO) are far in excess of acceptable levels.

In 2017, an Indo-Dutch collaboration, LOTUS HR Project, for cleaning and beautification of the Barapullah Drain was launched in the presence of dignitaries. Two years later, the second phase of the project – capable of treating about 10 percent of the discharges flowing through the drain – was inaugurated by King Willem-Alexander and Queen Maxima of the Netherlands. Yet, in June 2024, a submission before the National Green Tribunal of India (Execution Application No. 5 of 2024) argued that the cleaning process of the drain is affected considerably by the pillars of an elevated road built over a significant part of its length, also partially blocking the storm water flow. On the other hand, in January 2025, the Department of Environment, Govt. of NCT of Delhi in its report on the Progress in Rejuvenation of River Yamuna Against Action Plans and Timelines, has claimed that 91 percent physical progress has been made in tapping the Barapullah waters. The same report suggests that over 500 unauthorised colonies do not have sewer lines serviced by sewage treatment plants; we can only surmise where that sewage lands up. Clearly, there are contradictions in the report.

It does not make sense to point a finger at just industries or government agencies, though.

The majority of citizens in India are oblivious of the fact that their own excess consumption is a key contributor to the steady decline in water quality as well as availability. It takes about 150,000 litres (150 KL) of water on an average to produce one domestic passenger vehicle in India, including all its components. That's roughly the same amount of water that would be required to fulfil all daily needs of about 300 families. 2.7 KL of water goes into making a single cotton T-shirt. Just a tad less – about 2 KL litres – is the estimated water footprint for production of 1 kg of sugar. Leather goods involve substantial tannery waste production and ecosystem impact; towns like Kanpur and Kolkata bear the brunt. A pair of boots will use up 12 KL of water, and a leather tote bag 17 KL. Approximately 70 KL of water might be needed to construct a one-bedroom house in Bangalore. That's the same water footprint as that of a leather bag, a pair of boots and 15 T-shirts!

When educated Indians like me consume packaged water supplied in plastic bottles that are generating microplastics that find their way back into the food chain through water bodies, or use unnecessary high-SPF sunscreens that wash down to the sea and bleach exquisite coral reefs, they do not consider the broader consequences. Not necessarily because of lack of awareness, but because the consequences are not in our own backyard.

This means that every time we buy something we do not need, we are impacting water resources. Multiply that by 1.5 billion of us. When affluent city-dwellers leave the tap running while brushing their teeth or taking a shower, they may be indirectly forcing a villager dependent on daily wages in some remote, water-stressed part of the country to instead spend precious income-generating hours walking farther afield to collect two buckets of water for the family.

Ultimately, therefore, it is our individual behaviour and our collective will as citizens that will determine what legacy we leave behind. If we do not curtail consumption, and become frugal in our use of common resources, if we do not collectively hold industries and governments accountable every time and at every level, future generations will suffer the consequences of our irresponsibility.

CHAPTER 11

Water – A Journalist's Perspective

Hridayesh Joshi

"I discovered the secret of the sea in meditation upon a dewdrop."
~Kahlil Gibran

Hridayesh Joshi

Hridayesh Joshi is an award-winning journalist, writer and photographer. Three times winner of the Ramnath Goenka Award for Excellence in Journalism, he has received the National Water Award (Rajat ki Boonden) by NEER Foundation.

As a journalist, he has reported on a wide range of issues including politics, agriculture, health, rural affairs and internal security. He has written widely on current affairs, climate change, environment and energy. He has produced videos, reports and documentaries highlighting these.

Author of the book *Rage of the River – The Untold Story of the Kedarnath Disaster,* which is also written in Hindi, he has written a novel *Lal Lakeer (The Red Line)*. His translation of Bill Aitken's book, *Footloose in the Himalaya* is called *Aitken ka Himalaya.* His most recent work, *Samne Aate Hi Nahin* is the translation of Ramachandra Guha's *The Cooking of Books : A Literary Memoir.*

Water – A Journalist's Perspective

Hridayesh Joshi

It is almost impossible for any journalist not to write about water at some point in their life. The omnipresence of the molecule in our daily life, as well as its abundance and its scarcity, makes it a unique oxymoron. As a reporter, I encountered water very early in my career. What now resurfaces are indistinct memories, the grim stories of non-availability or of contamination of water – both in rural and urban parts of the country.

My first memories of water are not vague, but vivid; mostly cheerful. I remember how water was preserved in my village. When the rainwater would fall from the slanted roof, we would rush to bring buckets and pots and pans to place below the gushing stream of water before it fell to the ground. The children assisted in the mission of collecting rainwater and storing it in the kitchen or courtyard. It was as much a visual memory, as it was an acoustic treat.

The sound of water on metal utensils, changing tenor as pots and pans filled was a musical composition. The initial days of *chaumasa,* or the monsoon months, remain a time of experienced joy.

In those days, in the rural landscape I grew up in, there was enough water, both above and below the surface. Be it ponds, *naulas* (spring wells), falls or *guls* (small canals) in the fields, water was ubiquitous. There was no water supply, so women and men, and at times children as well, would fetch water for home use. But it never seemed a problem to us.

As I left the village to pursue my education and moved into the rented 'quarters' of urban *mohallas,* or neighbourhoods, water became a commodity, with a time-bound supply and there was a discipline that had to be observed.

I learnt about several aspects of water. I clearly remember, maybe at the age of seven or eight, when I first noticed a foul odour in tap water, I was told by my brother that some 'medicine', perhaps chlorine, had been added to make it clean and safe for consumption.

Growing up, cinema showed another facet of water. It was now used to compose aesthetic frames woven with the beauty of nature against the backdrop of compelling music. Directors portrayed romantic and sensual scenes of song and dance through rain-drenched visuals.

But in real life the absence and abundance of water has created serious problems. If a long wait for rain leaves a land parched and uncultivated, it plunges farmers into deep debt, even pushing them to the verge of suicide. A sudden flood ravages the countryside and inundates cultivated fields and ruins them.

Civilisation has had a complex relationship with water, that sometimes graces us with blessings or shows us its fury. Here are some stories from my reservoir of memories, which brought both challenge and melancholia.

The first is from my home state, Uttarakhand. Just a few kilometres from Pithoragarh, one of the biggest towns in the Kumaon hills, I met Sangeeta Devi, a woman in her late twenties. The radiant smile she wore on her face matched the elan she exuded when she took me to the water tank situated amidst a thickly forested hill. “It is from here we get our regular water supply for laundry and for bathing our cattle. Life is much easier now,” she told me.

The 400,000-litre tank was built by the local community in a small village of Gangolihaat block to store rainwater. I could not believe that in this place, situated just a few kilometres from the village where I spent my childhood, with no shortage of water back then, people now had to build a big tank to store rainwater. The change had arrived in just a span of four decades. But it was also a happy story where people understood that nature gives amply if you preserve it.

“Apart from building the tank to store rainwater, the villagers also started taking care of the jungle around it to replenish the springs in the hills that had been depleted,” said Rajendra Singh Bisht, who has been making villagers aware about water conservation for decades. “We planted trees like oak, rhododendron and *kaafal* (a wild berry) to increase moisture in the soil. As the health of our forests improved, the water output of the springs increased”.

Singh showed me how the villagers, mostly women, made trenches in the jungle, known as *Chaal* and *Khaal,* that store water and also prevent forest fires. However, scientific knowledge is required to identify the correct places to make trenches. Experts have helped villagers with this. After studying the hydrogeology on the slopes of Uttarakhand, scientists studied rocks’ behaviour, dipping, direction, and porosity of rocks to identify the place where water sources may have existed in the past.

Rajesh Joshi of G.B. Pant National Institute of Himalayan Environment at Almora explained hydro-geomorphological study. “The behaviour of rocks tells us about the possibility of water in that area in the past. Once we get this indication, we talk to the local community to cross check our scientific findings," he said. "If people confirm the existence of water sources in the past, we help them to reach the exact location of the origin." It shows how the local community and scientific community can work together to yield results.

Another lesson that I learned early in my career was how forests provide life to rivers. Be it Bhagirathi, Mandakini or Alaknanda (all tributaries of Ganga) or Yamuna and Tons (the biggest tributary of Yamuna), these mighty rivers originate from glaciers and gather water from the forests they pass through. Innumerable water streams and falls add to their power and flow as they proceed on their journey. As a child, or even as an adolescent boy, this had not occurred to me, though I saw so many springs and falls in the jungles around my village. Later, while reporting from Central India and south of Vindhya hills, I saw the flow of mighty watercourses like Narmada, Kaveri and Mahanadi. These rivers do not originate from glaciers. They emerge from springs in forests and their life and grandeur comes from the thousands of water streams that join them as they flow towards the sea.

Some believe that rivers flowing to the sea is a waste of water: why allow them to flow towards the coast and waste the water? We must use the water and thus interlink the rivers and build large dams on them. Today, there are more than one hundred large and small hydropower projects operating on the rivers of Uttarakhand alone. At least the same number of additional projects are planned on these rivers. We obstruct the flow of our rivers and submerge everything around, even if it displaces people, destroys forests and ecology, destabilises mountains and kills rivers.

We call it green energy, which reflects in our Nationally Determined Contributions or NDCs, the climate goals pledged by countries at United Nations climate talks. The power generation capacity of the proposed Pancheshwar Hydro Power Project in Uttarakhand is 5040 MW, which is more than the total capacity of all the existing dams in the state. If this project comes to fruition, it will be the biggest hydropower dam of South Asia. Besides the Himalayan slopes, Central India also has numerous dams. The Narmada Valley alone has over 30 large and small hydropower dams operational or under construction.

Why don't we allow our rivers to flow uninterruptedly? Someone who dedicated his life to rivers explained the importance of the flow of rivers to me. When we talk of rivers, many of us remember forester Manoj Misra, who unfortunately left us too early and too suddenly. He would often say. "Always remember that a river needs flow to keep it alive and clean. You take away the flow, the river will die. You provide flow to the river and the whole ecosystem around it will come alive."

Misra was one of the few activists who understood riverine ecology and fought to save it. His love and zeal to keep the Yamuna alive throughout its course, including in the national capital Delhi, where it receives most of its pollution, was remarkable. Every monsoon he would tell me: "Look at the Yamuna now. How different it looks when the stagnated stream receives water. The first thing to make a river clean, is to provide it with flow."

As the population has risen, the dependence on rivers has increased manifold. Not just for drinking water, irrigation or industrial purposes, but also for hydroelectric power. There are scores of dams on rivers which impact their flow. Some 200 km upstream of Delhi and just a few kms away from its origin at Dak Pathar near Dehradun, all the water of the Yamuna is diverted into the enormous Shakti canal. The river gets almost dry and its ecology dies. The death of the river is inevitable if its water is snatched from it. In the case of Yamuna, and many other rivers, the authorities hardly release any water into the river, except in monsoon when they are compelled to do so.

As if this is not damage enough, our riverbeds face huge construction and mining activity, all of which is almost unregulated. On one hand, the construction of flyovers, residential, commercial and religious buildings, railway bridges and roads encroach the river bed; on the other hand, unabated extraction of sand kills the river completely.

Sand is the second most extracted mineral or natural resource on earth after water, but policymakers do not realise that water and sand are different aspects of the river. Sand not only provides stability to the river system, it also facilitates habitats for various life forms, like nesting sites to crocodiles, turtles and birds. It is the sand in the river that ensures that we have groundwater.

In the second half of the twentieth century, when India opened its economy and the real estate boom followed, miners and builders led an assault on our rivers. The problem existed earlier too, but extraction has since become rampant and unsustainable. Government officials turn a blind eye to the illegal mining, sometimes even hobnobbing with perpetrators.

Many people living in cities do not realise how important sand is for the river. They regard opposition to sand mining, or in fact any kind of mining, or even demanding a sustainable extraction of minerals, as 'anti-development' activity.

With rapid urbanisation, as a section of humanity consumes water unsustainably, mostly for luxury, the water crisis is not surprising. Some people have the privilege of private swimming pools and water parks and fountains in their lavish apartments and five-star hotels but others walk miles every day to get a few buckets of water. They have no tap water supply at home. Such uneven and unfair distribution of this invaluable resource is a gross human rights violation.

India is a large country supporting almost 18% of the world population, yet it has only about 4% of water resources, disproportionately used by the rich and privileged. When surface water was not available easily, the dependence on ground water increased. During the 1950s and 1960s, the extraction of groundwater started rising, initially for drinking and then for other domestic needs and irrigation. As the pressure of population increased, the scale of extraction and consumption of groundwater also mounted rapidly.

This brought several problems in its wake. In India, as in many countries of the world, diseases caused by groundwater are common. Fluoride, arsenic, nitrate, chemical fertilisers and pesticides, as well as heavy metals have contaminated the groundwater. India is abysmally low on the global water quality index. In 2024, it was 120 among 124 countries.

In rural India, many people fall sick from consumption of contaminated groundwater. The number is overwhelming; yet governments seem incapable of handling the situation, simply leaving the people to their fate. The impact of contaminated water on the human body is often slow and governments are busy addressing 'urgent' problems.

Take the example of arsenic contamination which is a geogenic problem. In 2021, I travelled from Punjab to West Bengal, covering scores of villages and meeting hundreds of people who are victims of arsenic contamination in water. When I was covering the impact of contaminated groundwater on human lives, I met a doctor at Kolkata. When I asked him why so many people are sick and the problem of groundwater contamination remains unaddressed, he said the issue has to be 'very urgent' to be addressed. Government babus have no time for secondary issues.

"Try to understand. A file goes to a government office. They think that if they do this work even after a month, it is fine. Deal with an emergency first. While dealing with the emergency, when the time for this file comes, another emergency arrives and that problem remains unattended," he said.

Today this 'not so urgent' problem for the government has affected millions of people in India. In 2021, a study by IIT Kharagpur shows that more than 20% in area and over 250 million people in India, are exposed to toxic levels of arsenic in water. But most of them are forced to use contaminated water because they have no other option.

Saurabh Singh works with an NGO, Inner Voice Foundation. He took me to several villages where I saw people at the verge of death, due to arsenicosis. While filming a documentary I met one gentleman, Tarkeshwar Tiwari, who was suffering from the disease. He gave me a powerful interview narrating the woeful condition of the folks of invisible India who suffer silently. Tiwari was keen to see the film I was producing but did not survive the three months it took to complete the documentary.

Young students of Uttar Pradesh and Bihar are drinking water containing about 500-1000 PPB of arsenic, while the safe level is only 10 PPB. Singh raised a valid question. He said, "If they continue to drink such contaminated water, then a day will come when their kidneys, heart and lungs will get damaged. People often say that North India has a lot of population and human resources for paramilitary forces, defence and police - maybe we will not get any from here."

A few kilometres away from Patna, Deshraj told me this problem is denying jobs to youth in the armed forces. "We pass the written and physical examinations, but everyone is declared unfit during medical examination because of skin disease. The doctor tells us, 'When you change the water, you will be fit'. But this is the only water we have been using since birth. There has been no change and there is a lot of arsenic in it. Due to this about 90% of the youth of our village get filtered out. Only because of skin disease, caused by water. The remaining 10%, who get selected, are from wealthy homes. They have their own boring installed in their houses, which is 250 or 300 feet deep. Those people are fit. But people like us who are financially weak, do not have the resources to install personal hand pumps".

In Rajasthan, one of the most notorious states for stone cutting and mining in India, poor workers fall in the trap of silicosis, a terminal lung illness which is caused by inhaling silica dust over prolonged periods of time. Most of the victims who go to these mines or stone cutting factories are Dalits and Adivasis. The importance of water can be understood by seeing the large number of 'widow villages' and youth lying on their deathbeds. The lives of some of these people could have been saved if water was available for agriculture. Due to the water crisis, farming is not successful in these areas and the youth are forced to walk into this 'death trap' for employment.

This is not the story of one village, one district or one state. The crisis runs deep and wide. But there is also hope. People have demanded clean water from the government and found ways to solve the problem of scarcity. I have covered stories from Uttarakhand to Rajasthan to Madhya Pradesh where communities have fought against the odds in various degrees.

I have given examples of how people in Uttarakhand made successful efforts for water conservation and rejuvenation of water sources. Such examples are not limited to any one district or state of the country. Many positive changes have been seen due to community efforts in the areas of Rajasthan where people are craving for water or are falling prey to diseases like silicosis.

In Karauli district, which is affected by silica mining, the villagers created several large and small ponds with their hard work, and solved the water problem. An NGO, called Tarun Bharat Sangh, organised communities to dig hundreds of ponds across Rajasthan. There is no dearth of government schemes but corruption and administrative negligence has been a problem. Wherever the administration is proactive, there have been positive results.

My journalistic experiences related to water are too diverse to write in a single essay or probably even in a book. There are challenges and situations where one feels melancholic and frustrated but the stories of success keep hope alive. A scientific definition may term water a colourless and odourless liquid, but a rainbow of several colours around it offers an altogether different perspective.

CHAPTER 12

Water Future in a Climate Risked World

Reimaging a New Paradigm of Sustainable Management

Sunita Narain

"We can be water-secure if we choose to be water-wise."
~Sunita Narain

Sunita Narain

Sunita Narain is a renowned environmentalist, writer and the Director-General of the Centre for Science and Environment (CSE), New Delhi. She also edits *Down To Earth,* a leading environmental magazine. For over four decades, she has shaped public policy and opinion on critical environmental challenges, especially in the Global South.

Narain champions a science-based, pro-poor approach to environmental issues, advocating for cleaner air, water conservation, pesticide regulation, and sustainable urban development. She emphasises that environmental sustainability must go hand in hand with inclusive and equitable growth.

Her contributions have been widely recognized, earning her the Padma Shri, Stockholm Water Prize, Edinburgh Medal, and the Order of the Polar Star from Sweden. She has served on the Prime Minister's Council for Climate Change, chaired the Tiger Task Force, and contributed to the National Ganga River Basin Authority.

Named among the world's top 100 public intellectuals and featured in *Time* magazine's list of 100 most influential people, Narain's work continues to inspire efforts toward a more sustainable and resilient future.

Water Future in a Climate Risked World

Reimaging a New Paradigm of Sustainable Management

Sunita Narain

In this decade we will see the revenge of nature as climate change impacts increase. We need to scale our work to invest in local water systems to capture every drop of rain, so that we can build local resilience against drought. We also need to do it in our cities – lakes and ponds are the sponges that will allow us to harvest the rain-flood and make sure it does not become wasted water. We need to protect our forests and green spaces, as this is how water recharge will increase. In these times of water stress, not only must we make sure that wastewater or sewage is treated, but that it is recycled and reused. The waterbodies that we protect in our cities – the same ponds and tanks that we use to divert and harvest rainwater – can be used to channelise the treated sewage, and in turn, recharge groundwater. Water to water. We can be water-secure if we are water-wise.

In the past few years, climate records have tumbled like never before. First, the planet observed its hottest June on record. Then came July, which recorded the hottest two weeks ever. In June, the surface temperature of the world's oceans was the highest ever; Antarctica's ice was the thinnest ever. The heat has gone off the charts in many parts of the world. Never before have there been extreme rainfall events that have resulted in such large-scale loss of property and lives. Scientists warn that this is only the beginning of our season of despair, because much is changing in the natural world around us.

The unprecedented intense heatwaves in the oceans, including in the Bay of Bengal, are changing the way the monsoon winds blow over the Indian subcontinent's landmass. The western disturbances, which emanate from the Mediterranean region and bring snow and rainfall to the subcontinent in the winter season, are becoming more variable. This, scientists believe, is linked to temperature changes in the Arctic. In 2023, and again in 2024, the western disturbances that came as late as in June, collided with the monsoons and resulted in extreme rainfall, floods and devastation. This is now going to be combined with the much-warmer-than-anticipated North Atlantic Ocean temperatures, which drive extreme weather disturbances like cyclones, and the onset of El Niño in the Pacific, which is a precursor of extreme heat and drought. The bottom line is that there is no weather pattern for scientists to follow. We are out of our comfort zone and will be exposed to nature's dictates like never before.

Many parts of the world that are heating up, will need to find solutions for thermal comfort - including installing air conditioners. What this will mean for our warming world will depend on what these countries do to reduce the emission intensity of their energy grid. It will also depend on their ability to do what has been discussed for long; retrofit their buildings for better insulation and for ventilation. The fast-warming world will require inventive solutions so that we do not end up blowing up even more of the non-existent carbon budget.

We have millions of farmers, who are the worst impacted by extreme and variable weather. The situation is so bad that they have no idea when to plant their next crop. The planting season, which starts with the advent of the monsoon, has come and gone. What they plant is devastated by flood or by lack of rain. There is no crop calendar that they can work with now; and their risks are compounded many times. This is the tragic human face of extreme weather that we are seeing over and over again. Let's not lose sight of this.

We are oblivious of the need for climate-resilient development. In fact, we are doing the reverse. We are building on the floodplains, choking drainage and filling up waterbodies. We are mindless about the fact that the Himalayas are the world's youngest mountain range and so, development of everything from buildings to roads and other infrastructure must be based on this vulnerability. The devastation in our world is not only because of climate change, as our politicians would like us to believe. Climate change is exacerbating our human-made mess. I say this because we must not slip into believing that fixing this mess is beyond our control. Then there is no way ahead. This would be the ultimate tragedy.

We must understand that there are two incontrovertible facts: water is a key determinant for health security and economic growth. Secondly, water wars are not inevitable, but will happen only if we do not manage our resources prudently. I would also argue that while water scarcity is indeed growing, it is not inevitable that cities will run out of water or that we will not have any water to drink. I say this because water is a replenishable resource - it snows and rains each year. More importantly, other than in the case of agriculture, we don't consume water. We use and discharge. Therefore, it can be treated and then re-used and recycled. This is one future we can change.

This means getting the policy and practice of water management right. The good news is water literacy has grown. Over the past decades the country has learnt critical lessons on water management and evolved a new paradigm. Till the late 1980s, water management was largely confined to the issue of irrigation projects – building of dams and canals to store and supply water long distances. But then came the big droughts of late 1980s when it became clear that it was not enough to plan on augmenting water only through large projects. This was when the Centre for Science and Environment (CSE) published its report, *Dying Wisdom*, which documented traditional technologies for rainwater harvesting in ecologically diverse regions of India. The slogan was rain is decentralised, so is the demand for water. Therefore, capture rain when and where it falls.

There was a paradigm shift in policy. In the droughts of late 1990s, state governments launched massive programmes to capture rainwater by building ponds, digging tanks and building check-dams on streams. By mid-2000, these efforts coalesced into the Mahatma Gandhi National Rural Employment Guarantee Act (MGNREGA) – investing labour in building rural water assets. By this time, it was also understood that groundwater – considered a 'minor' resource was the 'major' supplier of water for both drinking water and irrigation in the country. It was also understood that over 50 per cent of agriculture was still rainfed and therefore, water conservation and decentralised rainwater harvesting was critical for productivity and wellbeing. This meant ensuring that every well and every waterbody was recharged.

In the decade of 2010, the crisis of urban drought hit homes. But again, policy evolved as it learnt that augmenting water supply was only one part of the challenge – cities were increasingly dependent on long distance sources; pumping and piping this water meant both losses in distribution as well as costs of electricity. This in turn made the available water expensive and more inequitable. As water supply diminished, people turned to groundwater. Without effective recharge, ponds and tanks had been decimated by real estate, or simply through neglect. This has meant declining water levels.

More importantly water supply was linked to pollution – the more the water supplied, the more is wastewater generation. This, without adequate treatment, leads to pollution of rivers and waterbodies, which in turn destroys available water and increases the cost of cleaning drinking water.

A few years later, research revealed that the bulk of urban residents are not even connected to an underground sewerage network, which is capital and resource intensive. Instead, they depend on on-site sewage 'disposal' systems, where household toilets are connected to septic tanks or just holding tanks or even to open drains in the vicinity. But the sewage treatment infrastructure was not designed to fit the city sanitation system and remained underutilised. Rivers remained polluted.

In all this, new solutions emerged. If affordable water supply was critical, cities needed to cut the length of their distribution pipelines, which meant increased focus on local water systems like ponds, tanks and rainwater harvesting. Then, if cities needed to ensure affordable sanitation for all and affordable treatment of wastewater, then on-site systems could be re-engineered so that waste was collected from each household, transported and treated. There was no need to build long distance pipelines for supply of water or even longer distance pipelines for taking back the wastewater for treatment. But most importantly, we have learnt that if this urban-industrial wastewater is treated for reuse then water is not lost. More importantly, our rivers will not be lost. This is where implementation is now focused.

The Indian experience has been invaluable in teaching, not just us, but the world how water management can be reinvented so that it is affordable and sustainable; it puts water in the hands of communities and focuses on decentralised recharge and reuse. Making water everybody's business is the only way ahead. This is because the key challenge is to ensure sustainability of water supply systems.

The government of India's ambitious and much needed *Har Ghar Jal* mission has recognised this fundamental flaw in water infrastructure projects and has stressed that its objective is sustainability so that water continues to flow in pipes and taps. This requires focus on improving the durability of the water asset that is created. It means ensuring that the pond or lake or tank is not encroached; and that the watershed that is critical for drainage to be secured is not destroyed.

The problem lies in the fact that land and water bureaucracies are fractured – somebody owns the pond; another agency the drain; and yet another the catchment. Water security implies a much-needed change in this prevalent system. This means giving greater control over the water structures to the local community. Deepening democracy and devolution of powers is then the answer to water management.

In all this we must minimise our use of water, become considerably more efficient with every drop. This means doing everything from investing in water efficient irrigation, household appliances and changing diets, so that the crops we eat are water prudent.

This is the opportunity and challenge. This decade we can put all we have learnt into practice and thereby turn around the water-story of India. It is possible. We just have to make it our single-biggest obsession. Water, remember, is about livelihood. It is about food and nutrition. It is about our future.

This is even more important in today's climate risked world. In this decade, we will see the revenge of nature as climate change impacts increase. We need to scale our work to invest in water systems and to make them durable, not just to withstand another rain, but another deluge. We need to speed up our work, because climate change will make sure that we have more rain, but in fewer rainy days. This means doing much more to capture the rain, when and where it falls so that groundwater is recharged.

This lesson is particularly important for the most ecologically vulnerable regions. The Himalaya are the world's youngest mountain ranges, this means that slopes are unstable, built on moraine – practically loose soil – and so prone to landslides and erosion. Combined with this, is the fact that this region is highly seismic – it is classified as the most earthquake prone. Then there is the added factor of climate change, which is bringing more volatility because of unseasonal and extreme rain events. The bottom-line is that this region is 'different' – these are not the plains of India that are situated on alluvial soil; nor the Indian peninsular region where there is hard rock, it is not even the slopes of the Alps, where mountains have aged.

You will say this is kindergarten knowledge – why do I need to tell you this. But I do, because we have completely lost the ability to plan for the ecological uniqueness of the region. We have built mindlessly in the name of development. This is not to say that this region should not see roads, infrastructure or housing. The point is how are these built, where and how many. It is about planning and about ensuring that these plans are implemented.

Take for instance, the question of hydropower plants. Each disaster has the same message: learn to respect the mountains before you build tunnels, roads and houses. The fact is that this region has the ability to generate clean power from its mighty rivers. The question is how many hydropower units should be built, and how will the damage from these built projects be minimised? This is where the faith of engineers is blind to the region's fragility. Or you could argue that they believe that all of nature is there to be dominated and re-engineered.

I saw this first hand in 2013, as a member of a high-level government committee to discuss these projects in the same region where the latest disaster has struck. The engineers had come up with a grand plan for some 9000 mw from 70-odd hydroelectric projects. According to their estimates, delivered with proud confidence, they would modify up to 80 per cent of the mighty Ganga and its tributaries in these moraine slopes through diversion in tunnels and thereafter, concrete structures and turbines for generating power. They told the committee that they had 'planned' the projects so that the river would be left with some 10 per cent flow through the lean season. In this back-to-back project design, the river would be re-engineered to be a drain. This was their plan for development.

The alternative that I suggested, which of course, was dismissed and relegated to a dissent note, was that the projects that had already been built should be redesigned by mimicking the river flow. This would mean that more energy would be generated when there was more water in the river and less during its lean times. At all times, the river would flow and not be re-trained according to their plan. We even showed, with the data that was provided by officials, that this would not result in loss of revenue for the projects. But it would mean that any future project would only come up if the river had excess water flow. It would mean many under construction or proposed projects, including the Tapovan-Vishnugad plant would be cancelled. This of course, was not convenient to the dam proponents. At that time I had written, and I am now again reiterating, that the engineers from the prestigious Roorkee University manipulated data to convince the committee that all was well. Over this last decade virtually all the 70-odd projects are now being built, disaster or no disaster.

The real issue is our inability to use available knowledge in policy making to ensure development is ecologically appropriate and socially inclusive. Why is it that all the institutions set up to design projects – and there are many located in this very region – are unable to provide this technical advice. Is it because their 'science' has become so fossilised and arrogant that they cannot comprehend the 'nature' of their business? Or is it because policy making has become so obstinate and obdurate that it refuses to entertain positions that are not in conformity with the chosen view? In this echo chamber there will be more disasters – let's at least accept this.

Our water future is about our water wisdom and in this we must recognise that water and culture go together. Water shortage is not about mere failure of rain. It is about the failure of society to live and share its water endowment. We can be water-secure if we are water-wise.

*“Water is life’s matter and matrix, mother and medium.
There is no life without water.”*

~Albert Szent-Györgyi

CHAPTER 13

How Dams Reshape Lives and Landscapes

Varsha Singh

"Have you also learned that secret from the river;
that there is no such thing as time."

~Hermann Hesse, Siddhartha

Varsha Singh

Varsha Singh is an independent journalist based in Dehradun, India, covering environment, climate change, and health issues. Since 2017, she has been reporting independently from eight Indian states. Her work has been featured in Down to Earth, BBC Hindi, Mongabay Hindi, IndiaSpend Hindi, The Third Pole, Oxpeckers, and other national and international media outlets.

She has received multiple fellowships and grants, including from Internews' Earth Journalism Network, the Journalism Centre on Global Trafficking, Climate Tracker, the Critical Action Centre in Movement (CACIM)'s Centre for Financial Accountability (CFA), and The Thakur Foundation. Her investigative reporting covers issues such as water and dams, groundwater, wildlife crime, just transition, biodiversity, and environmental crime.

Varsha's work has been recognized with The REACH Media Award (2023) for outstanding reporting on TB and a Jury Citation at the Laadli Media Awards (2021, 2022). She has also participated in international journalism workshops on climate change, renewable energy, and biodiversity.

In addition to journalism, she has contributed to national dailies such as Jansatta and Dainik Tribune.

How Dams Reshape Lives and Landscapes

Varsha Singh

Rivers begin their journey as vibrant lifelines, carving through valleys and plains, nurturing ecosystems and sustaining countless lives along their banks. Once free to flow and shape the land, they now find themselves shackled by towering dams, their natural rhythms disrupted and their vitality subdued. This confinement does not just threaten the rivers' existence; it endangers entire ecosystems: fragile aquatic life, diverse fish species, thriving aquaculture, and the livelihoods of millions who depend on them. From their origins to their meeting with the sea, the story of rivers has shifted from one of freedom and abundance to their struggle against human intervention.

Rivers flow through our lives in ways both subtle and profound. They are the lifeblood of our civilisation, nourishing lands and filling granaries that secure our food. Millions of Indians owe their survival to rivers: boatmen navigating their currents; fishermen casting nets; and communities gathering along their banks for prayer, rituals and solace. Rivers are more than geographical features, they are spiritual and cultural anchors, pivotal to festivals, folklore, stories and memories. They cradle our past, and their future will define the path of our survival.

But rivers are increasingly under threat. After independence, as the country's aspirations and needs grew, large-scale dam construction began to meet demands for water supply, irrigation and power generation. These development projects reshaped the geography of rivers and had long-term effects on their hydrological systems, altering their natural flow patterns and disrupting the ecosystems that depend on them.

Freshwater biodiversity has suffered immensely, as aquatic life depends on delicate balances, like flow patterns, water temperature, sediment levels, riverbed structure, which are disrupted by dams. This imbalance extends far beyond the riverbanks, affecting wetlands, floodplains and the millions of people whose lives are intertwined with these rivers.

Globally, freshwater biodiversity has seen the steepest decline of all habitats, with populations of migratory freshwater fish plummeting by 81% between 1970 and 2016. Scientists warn that dam construction, coupled with deforestation, climate change and intensive land use, is pushing ecosystems toward irreversible tipping points.

Rivers, which form the backbone of Earth's water cycle, are bearing the brunt of this crisis. Seven out of ten rivers worldwide face moderate to high threats, and over 80% of the global population lives in areas experiencing water stress or biodiversity risks. With more than 62,000 large dams and 4,200 hydropower projects reshaping rivers, nearly half of the world's rivers are moderately to severely impacted by flow alterations, habitat fragmentation and pollution.

India plays a significant role in this global story, ranking third after China and the United States in the number of large dams. According to the 2023 National Register of Large Dams, the country has 6,281 dams, including 143 under construction. While these structures have supported development, they have raised pressing concerns about sustainability.

Over the years, large dams have sparked widespread protests across India, drawing attention to the environmental and social costs they impose. From the higher Himalayan regions to the coastal plains, communities have risen in defiance, their fears rooted in the devastating consequences they have already witnessed - displacement, loss of livelihoods and ecological destruction.

One such case was in 2021, when the disaster caused by the avalanche and glacial lake outburst in Chamoli district of Uttarakhand was worsened by ongoing hydroelectric power projects. In my reporting, I visited the historic village of Reni, known for the Chipko movement, which was severely impacted by this catastrophe. The village was declared unsafe for habitation, forcing its tribal residents to seek new homes. Their access to land, forests and water, fundamental to their survival, was disrupted beyond repair.

The struggle continues in other parts of Uttarakhand. In Joshimath, local residents have long protested a large-scale hydropower project, fearing that the construction of tunnels beneath their town is contributing to the alarming land subsidence. Their homes and livelihoods are at risk, and they continue to voice their concerns as the situation worsens.

The impact of dam projects is not confined to the hills. This situation continues unabated, despite protests against large dam projects, such as the Narmada Bachao Andolan, which opposed the Sardar Sarovar Dam on the Narmada river due to concerns over displacement and environmental damage. Similarly, in Arunachal Pradesh, recent demonstrations have emerged against proposed hydropower projects on the Siang River, reflecting the ongoing resistance to large-scale dams and their consequences.

In Gujarat, the construction of the Bhadbhut Dam at the mouth of the Narmada river is threatening the livelihoods of thousands of fishermen. In 2024, while reporting from Bhadbhut village, I spoke to local fishermen who fear this may be the last generation to catch fish in the Narmada's waters. Since the Sardar Sarovar Dam was completed in 2017, water levels have steadily dropped, impacting fish populations. Fishermen's incomes have dwindled, and the approaching completion of the Bhadbhut Dam has deepened their sense of loss. Once the dam is finished, they fear that the river's remaining fish, their boats, and their way of life will disappear.

At Lohari village in Dehradun's Vikasnagar, the 120-megawatt Vyasi hydropower project turned into a nightmare for the locals, damming the free-flowing river. It also serves as a stark reminder of how the pursuit of green energy can cause significant harm to the environment, people and rivers.

The river that once flowed freely now struggles to find its course
The village that once sat by the river's bank now lies submerged within it.

The small village of Lohari, once located on the banks of the pristine Yamuna River in Dehradun's Vikasnagar, now exists only in memories. The river that once flowed freely through lush green mountains and dense forests is now dammed by the Vyasi hydropower project. The once-thriving community, intimately tied to the river, has lost its home, and the tribal culture of Jaunsar Bhabar celebrated here.

In the 1970s, villagers from Lohari, Lakhwad and other surrounding villages found themselves negotiating the fate of their land. Though many signed official documents, the decision was not made freely, as the government gave little room for negotiation. Over 40 villages were affected by the Lakhwad-Vyasi hydroelectric project. Public hearings were held. The villagers' collective stand was clear: "After giving up our land, we will not become landless." Compensation and land-for-land were the two main points included in the agreement.

The continuous flow of the Yamuna River, originating from the Yamunotri glacier, had been a symbol of prosperity, wealth and culture for this region. Locals revered the river, calling it 'Mother', and worshipped it. The river's waters nourished their fields, ensuring abundant harvests. The village granaries remained full and there was always enough fodder for the livestock.

The Yamuna's natural flow was soon reimagined as a source of power for the growing populations of Uttarakhand, Delhi, Uttar Pradesh, Himachal Pradesh, Haryana and Rajasthan. This vision gave rise to the Lakhwad-Vyasi hydroelectric project, conceived in 1979 to harness the river for electricity.

The 120 MW Vyasi project, one of the key components of the plan, would affect six villages with 336 families, including Lohari with 72 families, directly dependent on the river and its resources for their livelihoods.

Construction of dams, powerhouses and a barrage, began in 1987, overlooking the environmental and social risks of damming this vital water source. Work was halted in 1992 due to a lack of funds, the project's revival lingered, leaving riverside communities in prolonged uncertainty, fearing the loss of their land and the Yamuna's natural flow.

The risks of tampering with river systems became tragically clear in 2013 when Uttarakhand faced one of its worst disasters. Excessive rainfall and the collapse of the Chorabari glacier triggered devastating floods and landslides. These calamities highlighted the fragility of the region and underscored how large hydropower projects exacerbate vulnerabilities. In the aftermath, the Supreme Court ordered the cancellation of several proposed hydroelectric projects, including those in the higher Himalayas, due to the growing concerns over their environmental and disaster-related impacts.

While a ban was placed on hydropower projects in the high Himalayan region, efforts continued to accelerate projects in lower Himalayan areas, including the Lakhwad-Vyasi project. This project resumed in 2014 without a new environmental assessment or proper disaster risk evaluation.

The risks of altering the natural flow of rivers were tragically reinforced once again in 2021, when an avalanche triggered a devastating flood that swept through the region. The Rishiganga and Tapovan hydropower projects, built along these powerful waterways, were among the worst affected. The flood caused by the sudden surge of water resulted in the destruction of the Rishi Ganga project and severe damage to the Tapovan project. Tragically, more than 200 lives were lost. This was a stark reminder of how damming rivers could lead to even more disastrous consequences for communities and the environment.

Generations in Lohari village lived off the fertile fields nourished by the Yamuna River. The river's waters provided for both the people and their animals, ensuring there was always enough food and fodder. Gullo Devi, who came to Lohari as a bride, sustained her family by working the land. In her 70s, she watched as her daughter-in-law tended to the fields, her young grandson too small to run through them.

By 2021, Gullo Devi and the entire village were losing sleep. Even in her dreams, she saw the water rising over her village and it frightened her. Her nights were restless. Gullo Devi would wake up in the middle of the night, saying, "The reservoir has reached our village."

The people of Lohari village began sifting through old documents, where their ancestors had once signed agreements. The words were clear: "We will not become landless." Yet, the reality they faced was the loss of their land.

The government offered compensation based on current market rates, but for the villagers, it was no remedy. "Resettle us together. Give us land for land. Provide us with space for our animals. Return our forest", they insisted.

As protests grew louder, Gullo Devi's daughter-in-law, Asha Chauhan, raised her voice in anguish, "It's not just our village that will be submerged in the reservoir. Our culture, our identity, our way of life - everything will drown. Where will we go?"

The local administration identified land nearby for resettlement, along with a few dilapidated government buildings. The villagers, however, were not convinced. They asked the officials, "How can we leave our homes, courtyards, communal spaces, livestock enclosures and forests to live in these two small rooms? Where will we keep our animals? Where will we get fodder for them? These fields, where we grow wheat, maize, ginger, chilies, onions, garlic, turmeric and lentils, are sinking. Where will we farm now?"

The villagers' pleas fell on deaf ears. On World Environment Day, June 5, 2021, the people began a protest demanding resettlement and land-for-land compensation. The protest continued until Gandhi Jayanti, October 2, 2021.

It was the season for paddy planting. Rajendra Singh Tomar said, "This is the time to work in our fields, yet we are here, protesting with our children. We are from a tribal region. The law itself says that Scheduled Castes and Scheduled Tribes should be given land-for-land compensation."

Neither Gullo Devi's voice was heard, nor Rajendra Singh

Tomar's protest acknowledged. At 4 a.m. on October 2, the police arrived and arrested the protesters, sending them to jail in Dehradun. Five days later, they were granted bail by the High Court. These were people who had taken a stand for their homeland.

Inside their wooden homes, as they prepared meals over the fire, or walked along the rocky village paths, gathered at the village square or animal shelters, and worked in the fields, one sound echoed through Lohari: "We don't want money, we want land-for-land. Settle our village in one place. In just two days of rain, the water level in the lake rises. This lake haunts us, even in our dreams".

As the Vyasi Lake steadily grew, it began to engulf Lohari village's mango orchards. The animal shelter, earlier adjacent to the village, now stood across the lake, lost to the rising waters. The bamboo bridge that had once connected them to their animals had vanished into the lake. Watermills had already been submerged, and the cremation ground was drowned.

To mark the lake's advance, a yellow line was drawn on a field in the village. Once the water reached 626 meters, their fields would be submerged; at 631 meters, the entire village would be lost. The traditional two-storey wooden houses would disappear beneath the water. The once-thriving riverbank community would be swallowed by the lake.

The rising water mirrored the tears in the eyes of the women of Lohari. Standing knee-deep in the water, they voiced their defiance against the dam and the loss of their homes.

In December 2021, the Vyasi hydropower project began a trial run of its 60-megawatt turbine. This marked a turning point for the Yamuna River. To generate electricity, the river's natural flow was repeatedly halted, diverting water to fill the reservoir, leaving downstream stretches nearly dry. This disruption altered the river's ecosystem drastically, leaving aquatic life vulnerable.

The diminished flow revealed the river's distress. Fish, unable to survive in the shallow, deoxygenated water, surfaced and became easy prey. "The river was nearly lifeless," said Yashpal Tomar, a member of a local committee focused on the Yamuna. "Fish writhed at the surface, struggling for life, while people used sticks and stones to kill them. The sight was a brutal reminder of what had been lost."

The drying riverbed exposed more than just dying fish. It highlighted the fragility of a system heavily dependent on steady water flow. Concerned residents documented these devastating changes, sharing images of the parched river and its dead aquatic life with Uttarakhand Jal Vidyut Nigam Limited (UJVNL) and the district administration.

Amidst rising public anger, UJVNL altered its operations. To pacify the locals, the dam's water-filling activities were slowed during the day. The practice continued at night.

On April 15, 2022, in the former school near the village, piles of salvaged wood - windows, doors, door frames and roof planks - lay scattered. These were the remnants of homes swallowed by the rising water. Across from the school, the Vyasi reservoir reflected the submerged village, with only a few taller houses protruding above the waterline, a fading reminder of what once was.

Families displaced from their ancestral village were scattered, forced to rebuild their lives elsewhere. Some sought shelter with relatives, others rented homes, while a few had prepared in advance for this eventuality.

The Vyasi reservoir, holding water at 630 meters for the hydropower project, had drowned the village and the villagers' sense of stability. Watching their homes slip beneath the water, tear-filled eyes mirrored the rising lake. Relatives returned, daughters who had moved away came to bid farewell to their ancestral homes, standing at the water's edge, mourning, "The courtyards where we laughed and played are now gone." It was a profound loss of a community uprooted, homes submerged, memories now part of the lake's depths.

Gullo Devi's household belongings were scattered in one of the rooms of the school. Sorting through them, Gullo Devi said, "Take a picture of us that reaches Modi-ji directly. Show the world the injustice we have endured. He came to Uttarakhand in December and inaugurated this project. Take a picture that carries our voice to him so we can have a place to call home." She wanted to highlight the contrast: the government spends crores on official visits while villagers who lose their lands are left to fend for themselves, dissatisfied with the compensation and still fighting for justice.

The fodder kept for the livestock, floated aimlessly on the rising water. "Take a picture of that, it's our livestock's food," Gullo Devi's words, laced with despair, drifted across the surface of the lake.

Amid the protests on April 5, 2022, the residents were served with a notice to vacate within 48 hours. They pleaded and asked the administration for more time. "They didn't even give us a week. We pulled out what we could: beddings, utensils. We tore down windows, doors and the roof. Now, it's all just a heap of wood. These used to be our homes," Rajendra Tomar said with tears in his eyes.

As the village sank, people from nearby villages and other areas came to witness the destruction. The submerged homes in Vyasi Lake briefly became the backdrop for selfies.

The Yamuna was dammed at a time when rivers around the world were being freed from dams to restore their natural flow and protect aquatic life. In the US, the removal of the Klamath River dams was helping to restore salmon habitats, while in Europe, a record 487 dams were removed in 2023 to ensure the ecological health of rivers and their surrounding environments. This stretch of the Yamuna was one of the last unthreatened habitats for the endangered Mahseer fish, which relied on free-flowing rivers for survival.

Some nations have recognised the legal rights of rivers. In 2017, the Uttarakhand High Court granted the Ganga and Yamuna the status of living entities, along with forests and grasslands, acknowledging their inherent rights. However, this groundbreaking ruling was contested by the state government, and the Supreme Court issued a stay on the High Court's decision. Had rivers in India been granted legal rights, there would be many battles in courts.

Large dams bring significant risks, including environmental hazards, disasters, and the threat to ecosystems. In the context of climate change, the danger of dams is heightened, as their impact on river flows and aquatic life is well-documented. Studies, including the WWF's Living Planet Report, highlight that dams have contributed to a significant decline in biodiversity, with many species suffering as a result.

Despite the environmental loss, experts do not entirely reject dams, viewing them as a necessary risk for power generation, agriculture and drinking water supply. However, the relentless push for more dams and the attempt to lock entire rivers into reservoirs cannot be the way forward.

The solution does not lie in building more dams, but in how we manage water resources. The late Manoj Mishra, a former civil servant and convener of the *Yamuna Jiye* campaign, devoted his life to the Yamuna river and once said, 'Better water management can solve water's greatest challenges.' For cities like Delhi, sustainable water management is crucial to ensure long-term water security for all, without over-relying on dams. Our future depends on effective water management that reduces dependency on dams, protects the rivers that sustain us, and ensures water for generations to come. It is up to us to decide the kind of future we want: a future where rivers continue to flow freely, supporting ecosystems and communities, or one where our growing demands remain unmet, despite an endless expansion of dams.

References

1. https://www.icold-cigb.org/GB/world_register/database_presentation.asp?utm_source=chatgpt.com
2. https://globalenergymonitor.org/projects/global-hydropower-tracker/?utm_source=chatgpt.com
3. https://wwflpr.awsassets.panda.org/downloads/2024-living-planet-report-a-system-in-peril.pdf
4. https://nidm.gov.in/pdf/pubs/ukd-p1.pdf
5. https://apnews.com/article/klamath-dam-removal-salmon-spawning-4240169b4bfa327a6a67383ab536e971
6. https://rewildingeurope.com/news/rewilding-europe-capital-supports-dam-removal-in-sweden/?utm_source=chatgpt.com
7. https://damsafety.cwc.gov.in/index.php?lang=en&page=homeinfo&origin=front-end&tp=1&rn=1#:~:text=India%20is%20home%20to%20some,in%20India%20through%20institutional%20strengthening
8. https://cea.nic.in/wp-content/uploads/hpm/2021/11/Vyasi_report_website_compressed.pdf
9. https://dialogue.earth/hi/4-hi/82369/
10. https://hindi.livelaw.in/2017/07/08/499.html

CHAPTER 14

Water Migration – Shoots of Hope

Roli Srivastava

"Our village, our lives are all a shoonya (zero) without water."

~Ramlal Jat,

Sarpanch, Biliya Panchayat, Rajasthan

Roli Srivastava

Roli Srivastava is a Mumbai-based journalist and founder of The Migration Story. She reports on climate change, just transition, gender and migration, focusing mainly on India's marginalised communities. She has worked with the Thomson Reuters Foundation as Climate and Just Transition Correspondent, and Indian newspapers including The Hindu where she was Deputy Editor and The Times of India where she was Editor, Special Projects. Her stories have won her Ulrich Schwabe Media Award 2024 for Excellent Reporting on Drug Safety, the Fetisov Journalism Award (2020), Laadli Media and Advertising Awards for Gender Sensitivity (2020 and 2024) and a nomination for Journalist of the Year in the Drum Online Media Awards (2020). She was on the shortlist of the Chameli Devi Jain Award (2024). She won the COP29 Cross-Border Energy Transition Reporting Fellowship (2024), Robert Bosch Stiftung Media Ambassadors Fellowship (2015), and has presented her stories at the Global Investigative Journalism Conference (2019, 2021). She is also a journalist trainer.

Water Migration – Shoots of Hope

Roli Srivastava

Kharkhara, I thought, had a dry ring to it, like the sound of leaves rustling. Or an old man clearing his parched, itchy throat. Or footsteps on the bone-dry farms of this deserted, drought-hit village.

Its few inhabitants - the elderly, the sick, the poorest - sat outside their homes, staring at a street lined with padlocked houses. A golden dust filled the air, specks of it had settled on their cheeks that they had not bothered to wipe off. A sense of ennui enveloped all existence in this village when I walked around it in early 2018.

Kharkhara is a village of those 'left behind', like countless villages across India where the young, able-bodied residents have migrated for work, as farm yields wilt under the weight of climate extremes and years of water-intense farming practices. In the case of Kharkhara, located in Balangir district of the eastern Indian state of Odisha - one of India's most climate vulnerable regions ravaged by repeated cyclones, droughts, and floods - villagers trade their freedom for cash advances from labour contractors to work in brick kilns, ending trapped in debt bondage. And yet, others follow them.

"There is no choice really", I was told by Mathura and Maheshwar, a couple in their fifties. Their children and relatives had migrated, but they stayed back, unsure whether they would be able to cope with the backbreaking work at a kiln. There was also the fear of life beyond the village they had spent their lives in. Staying back meant living an isolated life, but they chose loneliness over exposing themselves to unknown cities and hardships.

A few years later, in a village in Bundelkhand on the border of Uttar Pradesh and Madhya Pradesh, a farmer told me that unlike the others, he had not migrated from his village to the city, opting for daily wage labour to make ends meet. It was a difficult life, but he had clung to the familiar, and to a distant past. "*Jab hum chhote the toh baadal mandir ki disha se aate the. Ab pata nahi kahan gaye*" (when I was a child, the clouds came from the direction of the temple. Now I don't know where those clouds have disappeared), he said, as if talking about friends who have migrated, and no longer came to play with him.

Distress migration from water-starved rural India often lands workers in risky, exploitative jobs. "But it is those who are left behind who suffer the most", a senior labour rights and migration expert, Umi Daniel, told me after I returned from Kharkhara. His words stayed with me.

Among the few young people who had stayed back in Kharkhara was Kumar Gwal, 40. He stood at the centre of his farm, cracks running through the soil beneath his feet. He spoke about his dilemma – to accept 100,000 rupees from a labour contractor for his family of four to work in a brick kiln, or stay another year in the village to see 'if it rains'. Kharkhara's empty streets indicated that his optimism was not widely shared.

For years, the drought at Kharkhara had yielded a steady supply of workers to the labour market in Kantabanji, a busy town about 115 kms away. Labour contractors scout villages in the district, offering jobless farm workers and farmers a rare opportunity to earn, bringing them to the labour market from where they board trucks, buses or trains that ferry them to brick kilns.

Kharkhara gradually emptied out.

With each passing year, even the charm of *poush poornima,* had diminished. Until two decades ago, workers from the village toiling on farms, often trapped in debt bondage, or *halia* (the traditional system of offering cash advances to villagers before taking them to farms locally referred to as halia) would get a brief reprieve during this time. The festival was an annual break, when the rich farmers who hired the workers would clear their dues, offer them grains and make another cash advance to ensure they would return to the farms after the festivities, sweetening the deal for enslavement.

Kharkhara was always impoverished. Yet on the full moon night of January, when poush poornima is celebrated, it would come to life, the festivities stretching over a few days. Workers freed from the farms would join their families. The aroma of mutton, fish or chicken curry, *puri* and *chaula khiri* (rice pudding), *pitha,* and other treats wafted from kitchens.

Mathura and Maheshwar recalled their busy kitchen. Their house, now quiet and desolate, once burst with people, food and laughter.

Eight years later, I spoke to a local villager in Kharkhara who told me the migration pattern had remained unchanged. More than 60% of the villagers have in the past year migrated to the brick kilns of Andhra Pradesh, Telangana and Tamil Nadu, and return during the monsoon when kilns close.

Poush poornima comes and goes.

"Farming is like betting now. There is no guarantee of return", a villager said.

Agriculture is India's largest employer, supporting the livelihoods of 250 million farmers and informal labourers – but their work is getting harder as climate change makes farming difficult, pushing up debt, migration and suicides.

An estimated 55 million people globally are affected by droughts every year, according to the United Nations' health body[1]. The United Nations also warns[2] that more than half of the world's major aquifers are being depleted faster than they can be naturally replenished, with India listed among countries most at risk of groundwater plummeting to levels that existing wells cannot access. For farmers in India, an agrarian economy where farmlands have been wilting under the weight of droughts – with 1987, 2002, 2009, 2015 and 2016 marked as the cruellest, as per government data[3] – the loss of yield has eroded incomes in this largely monsoon-dependent nation.

A 2022 study[4] linked drought with the probability of a rural Indian household having "at least one temporary migrant member". The study predictably concluded, "the probability of temporary migration on account of drought is more severe among the socio-economically marginalised sections of the rural population compared to their better-off counterparts".

About one in three people in India have moved or considered moving because of weather-related disasters, noted a 2023 report by the Yale Program on Climate Change Communication[5]. The report cited extreme heat, droughts, sea-level rise, flooding, or others that had already led to 14% Indians migrating.

1. https://www.who.int/health-topics/drought#tab=tab_1
2. https://interconnectedrisks.org/summaries/2023-executive-summary
3. https://agriwelfare.gov.in/Documents/Updated Drought Manual_0.pdf
4. https://journals.plos.org/plosone/article?id=10.1371/journal.pone.0275449
5. https://climatecommunication.yale.edu/wp-content/uploads/2014/ 05/climate-change-indian-mind-2023.pdf

Migration data is scattered in India with varying estimates from researchers, government agencies and nonprofits, of inter-state migrant workers. Climate-linked migration is even more poorly mapped. But a study of this isolated, scattered data throws up a pattern of migration fuelled by water scarcity.

Take Balangir for example, where Kharkhara is located. More than half of the nearly 40,000 workers who migrated from Odisha in 2023, were from Balangir, which records hot and dry summer and highly erratic rainfall distribution of south–west monsoon rains, according to government observations. Between 1974 and 2017, the district[1] recorded 20 natural calamities, of which 16 were droughts. It even suffered three episodes of flooding during this period!

Droughts in India, in regions such as Balangir, have been exacerbated by water-intensive agriculture, with rural livelihoods in most parts of the country wedded to paddy yields. This water-guzzling crop, while providing food security to the country and incomes to farmers, has worsened droughts, making their impact harsher.

A quarter of India's total crop production is grown in watersheds that use more water than can be naturally replenished, according to a World Resources Institute study[2]. The study warns that as temperatures continue to warm, groundwater depletion in India will triple by 2080.

Mounting agrarian distress has for years claimed farmer lives, with droughts fuelling debt and mental anxiety.

While migration continues to be the way out from this web of debt and distress for many, shoots of hope have begun to emerge in the lives of those left behind, with women emerging as

1. https://balangir.odisha.gov.in/departments/agriculture
2. https://www.wri.org/insights/growing-water-risks-food-crops

the superheroes taking on the Herculean challenge of tackling the rising climate stress.

About 500 kms from Kharkhara, in Kaurnikala in Kendujhar district in Odisha, women are learning to break free from paddy. It is a mighty task in this village where paddy is the predominant crop on acres and acres of farmland, visible as you enter the village.

On an overcast afternoon in 2024, villagers sat huddled under a massive tree, learning how to make millet fritters, curiously watching a campaigner chop onions, curry leaves and chillies and mixing it all in a millet batter, and then dropping spoonfuls of it into a frying pan on a wood-fired stove.

Millet, a traditional but forgotten drought-resistant crop, is making a global comeback. It is having a revival of sorts across Indian states, with officials banking on the small-grained cereal – dismissed for decades in several parts of the country as *mota anaaj* (coarse cereal) or even animal feed – to turn an aggressive high tide of climate risks.

The women at the fritter workshop, the 'left behind' wives of migrants, are among India's new crop of millet farmers, the hardy crop being held as a solution to protect yields and restore farmer incomes. The women hope the crop will restore the health of their farms and families, after years of hybrid rice cultivation with chemical fertilisers and pesticides having stripped the soil of its nutrients, while also adding to planet-heating farm emissions. Consuming the toxic rice, they said, had brought in diseases of hypertension and diabetes in their once-healthy village. Worse, water scarcity has made paddy harvest unpredictable, pushing their sons and husbands out to work in alien cities or risky iron ore mines across the region.

If fluctuating weather has created water refugees out of farmers, it has inadvertently catapulted the status of women from farm workers to climate solutions champions in places like Kaurnikala in rural India. Most women work as farm labourers or unpaid workers on family-owned fields, constituting nearly three quarters of the agricultural labour workforce. Yet they own less than 13% of farmland[1], according to government data. They are rarely seen as farmers, the profession still conjuring up an image of axe-wielding men sweating it out on farms. Extreme weather warnings are sent on mobile phones, but fewer than half of rural women have access to one.

In recent years, however, government agencies and nonprofits have enlisted women to carry out climate-smart farming techniques since the men have migrated.

Driving through Mumbai's manic traffic, chauffeur Nutan Pathak hopes these sustainable solutions take root and show assured income, for migrants like him to return to their village. He migrated from his village in eastern Bihar more than 20 years ago, and said he will not be able to return until he feels assured of making a living income from farming. This is currently far-fetched, owing to fickle rains impacting yields.

Reporting from villages over the years, I have found that beyond the cracks on fields, skeletal trees and farmers looking up into the sunny sky waiting for rain-laden clouds, are stories of resilience and solutions, and even of cottage industries that have emerged from the dry landscape.

Driving on the highway connecting Bhilwara and Udaipur in the northwestern state of Rajasthan, there is a sudden pop of colour on the roadside along the highway: mini trucks plastered with bright, colourful images of local gods and warrior kings, tall glasses of *lassi* and *falooda*, and ice cream scoops.

1. https://www.mospi.gov.in/sites/default/files/publication_reports/Annual Report, PLFS 2017-18_31052019.pdf

"It is the migration season", the driver told me in 2024 when I visited the districts of Bhilwara and Shahpura to report on a village that had seen migrants returning. But the bright ice cream trucks, even motorcycles remodelled into carts, loaded onto bigger trucks had my attention.

I was in Gangapur, India's unofficial hub for all things ice cream. On a kilometre-long stretch in the main market are nearly a dozen auto workshops where workers remodel Tata Ace tempos – or *chota haathi* as the vehicle is colloquially called - into ice cream carts, fitting refrigerators, shelves and neon lights. Shops sell refrigerators, others busily print posters for the carts. A bunch of new shops selling stainless steel ice cream ladles, chocolate and strawberry syrups, have sprung up in recent years.

In drought-hit Gangapur, an entire generation of ice cream migrants has emerged from decades of water scarcity. The ubiquitous 'Mewar ice cream trucks' on city roadsides have created job opportunities back home – each workshop hires 10 to 12 men. Locals estimate about 50,000 trucks are converted during the four-month 'season' – November to February - when the migrants return to meet their families and refurbish their carts.

The remittances from migrants have helped their families back home build concrete houses and install tube wells, both to irrigate their farms and for domestic use.

But it is not enough to bring the ice cream businessmen back home.

Over 600 miles away in Mumbai, ice cream seller Shankar Singh said there was no hope for water in his village, no future in farming, and the ice cream business was the only way forward. Until a few years ago, he owned a stationery shop in a Mumbai suburb, but saw a better opportunity in the dessert business, seeing his relatives and fellow villagers from Gangapur make a steady income in the forever hot and humid metropolis.

Singh stays in Mumbai and believes that if not for this business, he wouldn't be able to protect his family's future. "My family is able to eat because of this business. I can't close this shop even for a day", said Singh, who has overcome the drought-related debt and distress phase, but now eyes income to ensure his children get an education that will fetch them good 'white collar' jobs, so they don't have to spend months in an alien city away from their families.

The ice cream industry is a rare outcome of a drought, of a community's enterprise to turn a calamity on its head, turn distress migration into a business enterprise. I found Rajasthan had more such examples of communities creating solutions.

Rajasthan, India's largest state, is among the most vulnerable to droughts, with 98% of its 250 village blocks in sectors marked as 'dark zones' – areas with dangerously low groundwater levels – and almost 7% of the land uncultivable, according to officials at the Foundation for Ecological Security (FES), a land conservation organisation.

Rural Rajasthan has recorded out migration for years, but 150 kms from Gangapur, in Biliya panchayat of Shahpura district, men have started returning.

Biliya, a small panchayat of five villages, nurses big dreams. Large swathes of cracking, dry land, lie along the road leading to Biliya's villages. Families here have been torn away from each other for years by a relentless drought, with the men migrating to the textile mills of neighbouring Bhilwara district, others to states as far as Jammu & Kashmir and Maharashtra, to dig borewells. The mill migrants make 5000 rupees for 12-hour workdays at the mill, share a room with other migrant men and return home for festivals and marriages, their way of life for at least three decades now.

Farmers at heart, the mill workers believe they will lose their farms forever to a worsening climate crisis. They approached Ramlal Jat, Biliya's sarpanch, a few years ago to discuss a solution.

Now Ramlal Jat could have been a poet, a solutions expert even, in the global courts of climate discussions, if the couplets he spouted were Wordsworth's and not *Rahim ke dohe,* if his earthy wisdom and astute understanding of state funding had value in the shiny tomes of research on climate impact mitigation. I found Jat, who is trained as a lawyer, as much at ease in the corridors of government offices as in the newly-revived farms of his village.

My first meeting with him was on a nippy February morning, surrounded by villagers of Surajpura, one of the villages in Biliya panchayat, all of them sitting on a wicker mat with Jat, facing their lush fields, a thin smile on their lips. The sound of water gurgling into their farms, filled the air.

The men spoke over each other, with equal excitement and pride to tell the story of a wall they built, the trenches they dug and the anicuts they planned so rainwater reached their farms.

Biliya's villages suffered scarce rain in the sowing months of July to October, and unseasonal rainfall in winter that would damage its standing crops. Worse, its poor soil quality prevented water from percolating.

By the 1990s, a migration pattern had emerged, as the men realised there was no future in farming and started leaving. The women, as always, were left behind to tend to the children, the elderly and whatever was left of the farms.

"The circumstances were such. What could he have done," said Shanti Devi, drawing her sari over her head, covering her face, as she does in the company of men other than her husband

and sons. "If it doesn't rain, what work is there for my husband here?" she asked.

So Surajpura's villagers met Jat with a proposal to build a wall to arrest rainwater runoffs. Jat took it up as a mission. The villagers had initially built a four-feet wall that collapsed with the first spell of rain. They needed direction.

Jat tapped into the funds of Mahatma Gandhi National Rural Employment Guarantee Act (MGNREGA) and other state projects for groundwater resources management and rural development. He enlisted the support of local nonprofits working on natural resource conservation as well as the government, for technical guidance on building the wall.

On this morning, as the men spoke excitedly, the women were quiet, listening to them intently. It was only in the late evening that I found an opportunity to speak to the women alone. More than half the villagers who toiled on the wall for six months in 2022 were women.

This is not unusual. Women make up more than half of MGNREGA beneficiaries across India. Their participation rate is nearly 70% in states like Rajasthan, according to government data. This is largely because of the out-migration of men, researchers say, pointing at the massive rise in the scheme's uptake during the COVID lockdowns when migrants returned from cities.

In Surajpura, a vermillion-smeared rock placed under a tree, is widely revered as the wish-granting deity, Bhairon Dev. Saffron scarves tied to its branches carry their prayers for water, a silent record-keeper of the women's participation in the building of the wall, that indeed successfully arrested the rainwater, reviving farms.

A villager showed me two thin carrots and a few roses in an iron bucket that he had plucked from his farm, his excitement like that of a child showing off new toys. They spoke about the migratory birds such as painted storks that had begun to return. The women would share later in the evening the wins they scored: they would walk 5 kms to fetch water, but now with the wells recharged, the drudgery had reduced.

I continue to receive calls from sarpanch Jat, sharing that the wall survived a fresh spell of rain, that the trenches were full of water, that I should return to see how well their farms are doing, bringing back migrants. Jat said 60% of all migrants from the panchayat had returned.

Among those who returned was Mahavir Jat, 28, who had migrated to Jammu and dug borewells there for six years. He now leases a farm and rears two buffaloes, earning 3,000 rupees every day from selling their milk. Working in Jammu meant staying away from his family, missing out on his children's education and growth milestones, and missing his way of life in the village.

"I don't see the need to migrate now," he said.

CHAPTER 15

Monsoon and Water in a Changing Climate

Roxy Mathew Koll

"Everything depends on the rain."

~Matadin Meena,
75-year-old farmer from Rajasthan

Roxy Mathew Koll

Dr. Roxy Mathew Koll is a climate scientist at the Indian Institute of Tropical Meteorology (IITM), Ministry of Earth Sciences, Pune. He did his Ph.D. in Ocean and Atmospheric Dynamics from Hokkaido University, Japan.

Koll has made breakthrough contributions to the research, monitoring, and modeling of climate and extreme weather events over the Indo-Pacific region. His work has advanced the scientific understanding of monsoon floods and droughts, terrestrial and marine heatwaves and cyclones. His research has influenced state-to-national climate policies and disaster management strategies. He is a Lead Author of the IPCC Reports and the former Chair of the Indian Ocean Region Panel. He actively collaborates with citizen science networks, local governments and media to bring science to society.

Koll received the Rashtriya Vigyan Puraskar (National Science Award), the highest recognition in the field of science, technology and innovation in India in 2024. He was conferred a Fellow of the American Geophysical Union (AGU) and was awarded the AGU Devendra Lal Medal for outstanding research in Earth and Space Sciences in 2022.

Monsoon and Water in a Changing Climate

Roxy Mathew Koll

India's intimate relationship with the monsoon and water shapes its culture and economy. Today, climate change is altering our connection with water, with its impacts rippling through the monsoon, glaciers, rivers and oceans, affecting every citizen.

The geographical location of India, surrounded by tropical waters at one end and the Himalayas at the other, was once the region's forte, but no longer. The Indian Ocean is warming at the fastest rate and the Himalayan glaciers are melting rapidly. As a result, the subcontinent is witnessing a trend of rising floods, landslides, droughts, heatwaves and cyclones. This is affecting the food, water and energy security of the region, deeply endangering lives and livelihoods. India has become the poster child of climate change.

This chapter explores these interconnected challenges, highlighting water risks in a warming world while emphasising actionable solutions.

Escalating Water Stress

Globally, water systems are under unprecedented strain, exacerbated by rising temperatures that intensify evaporation, disrupt precipitation patterns and amplify extreme weather events. Currently, over a third of the global population faces water stress, a figure expected to reach half the population by 2050, as demand continues to outpace supply. This crisis is not merely one of scarcity but inequity, where the most vulnerable communities bear the heaviest burden. Urban centres dominate resource consumption, leaving rural populations, particularly in arid and semi-arid regions, grappling with limited access to clean water.

The plight of Denganmal, a village 140 kilometres east of Mumbai, is one of the cases that illustrates these disparities. Chronic water scarcity has driven households to adopt extraordinary measures, including polygamous marriages where 'water wives' are tasked with fetching water for their families. This stark reality highlights the human cost of water inequity, where women shoulder the disproportionate burden of climate change.

Monsoon Droughts and Floods

India's summer monsoon rainfall during June–September accounts for 78% of its annual rainfall, directly influencing agriculture, water resources and energy security. Every year from June to September, the monsoon delivers approximately 848 mm of rain across India, amounting to 2,700 trillion litres of water—enough to provide nearly 2 million litres per person. About half of this rainwater enters rivers, with the Ganges-Brahmaputra basin alone carrying 37 million litres per second. This extraordinary natural phenomenon shapes the contours of life and livelihood in India, but it is increasingly under strain.

Since the 1950s, the monsoon's behaviour has changed significantly. Prolonged dry spells now alternate with intense rainfall events, increasing the risk both of floods and droughts. While moderate rains spread through the monsoon season rejuvenate the groundwater, extreme rains flush water out easily into the sea or evaporate. This causes water insecurity, following flood events, within the same season.

Regional anomalies during the monsoon season have also become stark. For instance, in 2018, Kerala experienced catastrophic flooding due to excessive rainfall, displacing millions and resulting in economic losses exceeding $5 billion. Contrastingly, regions like Marathwada have witnessed persistent droughts, driving agrarian distress and migration.

India has experienced over 325 major flood events between 1950 and 2024, reportedly killing more than 81,000 people, displacing millions and causing reported economic losses exceeding $140 billion.

These floods are not confined to rural areas. Urban centres are becoming increasingly vulnerable. In Mumbai, urbanisation has intensified flooding, with impermeable surfaces preventing natural drainage. The floods of 2005, which brought the city to a standstill, are a stark reminder of the inadequacy of existing infrastructure to cope with extreme rainfall. Similarly, Chennai faces challenges where encroachments on wetlands and rivers exacerbate flood risks.

Figure 1(a) Trend in mean summer monsoon (June–September) rainfall, during 1950–2024. Brown colours indicate a long-term decline in the total amount of rainfall while green colours indicate an increase in the rainfall amount.

Figure 1(b) Trend in the frequency of extreme rainfall events over the summer monsoon, during 1950–2024. Orange-red colours indicate a rise in the number of extreme rainfall events while blue colours indicate a decrease in extreme rainfall events.

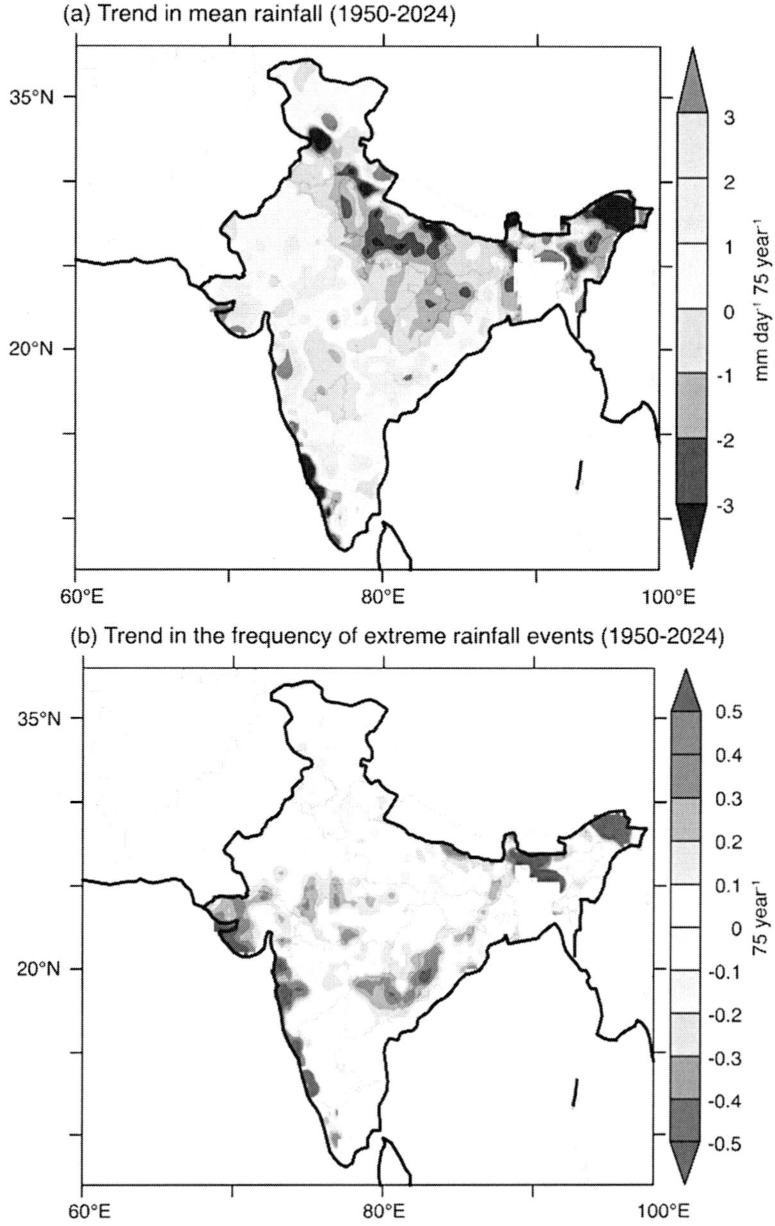
(a) Trend in mean rainfall (1950-2024)
35°N
20°N
60°E
80°E
100°E
3
2
1
0
-1
-2
-3
mm day⁻¹ 75 year⁻¹
(b) Trend in the frequency of extreme rainfall events (1950-2024)
35°N
20°N
60°E
80°E
100°E
0.5
0.4
0.3
0.2
0.1
0
-0.1
-0.2
-0.3
-0.4
-0.5
75 year⁻¹

Data from the India Meteorological Department shows that while the total amount of rainfall during the monsoon season shows a decline over several parts of India including the north and the Western Ghats (brown colours in Figure 1a), the core monsoon regions over India have witnessed a threefold rise in extreme rainfall events (red colours in Figure 1b).

Climate models suggest that extreme rain events in India could further increase multifold by the end of the 21st century, under high greenhouse gas emission scenarios. The implications of these changes are profound, with cascading effects on agriculture, water management and disaster resilience.

Melting Himalayan Glaciers

The Himalayan glaciers, often referred to as Asia's water towers, are retreating at a rapid rate. Over the past four decades, glacial mass loss in the Himalayas has doubled compared to the 20th century. This trend threatens perennial rivers like the Ganges, Brahmaputra and Indus, which depend on glacial melt for dry-season flow. The ramifications are far-reaching, affecting agriculture, hydropower generation and drinking water supplies.

Seasonal variations are becoming pronounced. Glacial melt contributes to higher river discharge in winter, which temporarily benefits downstream communities. However, as glaciers shrink, less meltwater is available during the critical summer months, impacting agriculture and hydropower. This dynamic is evident in regions like Himachal Pradesh, where hydroelectric projects face decreasing water flow, undermining energy security. Moreover, glacial retreat increases the risk of glacial lake outburst floods, sudden and catastrophic events that can devastate downstream areas. An example is the glacial burst that led to massive flash floods in Uttarakhand's Chamoli district in 2021, leading to 234 reported deaths.

Addressing these challenges calls for a multipronged approach, including enhanced monitoring of glaciers, early warning systems and investment in sustainable energy alternatives.

Coastal Vulnerabilities

Rising sea levels lead to saltwater intrusion and coastal erosion, and communities along India's 7,500-kilometre coastline face multiple risks. Increasing cyclones, storm surges and floods due to heavy monsoon rains heighten the threat.

Rapid ocean warming plays a major role in rising sea levels through thermal expansion of water. The warm waters also provide a consistent supply of heat and moisture to the atmosphere, intensifying cyclones and monsoon wind bursts, leading to severe storm surges.

Coastal cities are at heightened risk when these extreme events overlap, causing compound extreme events.

Cyclone Yaas in May 2021 exemplifies this phenomenon, where cyclonic storm surge, heavy rains and sea level rise combined with a high tide inundated the coastlines of Odisha, West Bengal and Bangladesh, several kilometres inland. Such 'compound floods' have the potential to submerge large parts of coastal cities for days, if not weeks. The saltwater intrusion damages agriculture for years and contaminates freshwater, threatening food and water security in coastal districts.

Adaptation strategies must be holistic, integrating local knowledge with scientific advancements. For coastal communities, this could mean creating early warning systems, developing resilient infrastructure and implementing ecosystem-based adaptation programs.

Mangroves have been shown to mitigate the impact of strong cyclone winds and demonstrate an ability to recover after sustaining damage. They also serve as natural floodplains, absorbing floodwaters caused by cyclonic storm surges. However, mangroves may not function effectively as a defence in all coastal regions. More research is needed to identify optimal solutions, likely involving a combination of natural and engineered defences, to provide comprehensive protection for vulnerable coastal areas.

Public Health Implications

Seawater inundation along the coast is linked to an increased risk of cholera outbreaks, as the bacteria responsible for cholera, Vibrio cholerae, often attach to zooplankton in coastal waters. When sea levels rise, the bacteria spread through flooding and contamination of freshwater sources in low-lying areas, leads to cholera outbreaks among coastal communities. In addition, cholera outbreaks often surge during monsoon seasons due to flooding and increased water contamination. A multifaceted approach that ensures water quality and sanitation, surveillance, social mobilisation, cholera vaccines and treatment is key to reducing cholera outbreaks.

Water insecurity arising from the climatic changes in the monsoon directly impacts public health, migration and social stability. Communities in drought-prone regions like Marathwada face crop failures, debt and distress migration. In contrast, flood-affected regions struggle with waterborne diseases and infrastructure loss. Dengue, a climate-sensitive vector-borne disease, illustrates the link between monsoon variations and health.

Studies show that warm temperatures and moderate rains during the monsoon season create optimal conditions for Aedes mosquito breeding, leading to an increase in dengue cases, whereas heavy rains can flush out breeding sites and reduce outbreaks. In

Pune, a dengue hotspot, mortality rates rise significantly during years with evenly spread moderate rainfall. Future projections suggest that rising temperatures and altered monsoon patterns could increase dengue-related mortalities in Pune by up to 112% by the end of the century under high-emission scenarios.

Dengue adaptation measures include utilising prediction models, promoting community-driven vector control programs and strengthening healthcare responses. Removing stagnant water and using larvicides are critical measures to curb breeding. Public health campaigns to raise awareness can further bolster these efforts.

River Interlinking and Climate Implications

River interlinking, proposed as a solution to manage India's water distribution imbalance, has sparked both optimism and controversy. The project envisions connecting major rivers in India through a network of 15,000 km of canals and 3,000 reservoirs, with the capacity to annually transfer 174 billion cubic metres of water. This ambitious USD 168 billion initiative aims to redistribute water from surplus to deficit basins, increasing the irrigated area by 30 million hectares. While it promises to mitigate droughts and reduce flood risks, emerging research highlights its potential ecological and climate implications.

Recent studies indicate that river interlinking can alter local climates, particularly the monsoon. According to the research, the proposed river interlinking projects could potentially lead to an increase in September rainfall of up to 10% in some parts of the Ganga, Godavari and Krishna basins while reducing the rainfall by up to 12% in several arid regions of the country, namely parts of Madhya Pradesh, Odisha, Chhattisgarh, Uttarakhand, Rajasthan and Gujarat.

Extensive interlinking affects regional hydrological cycles by altering evapotranspiration patterns and atmospheric moisture distribution, potentially weakening the monsoon system. Moreover, changes in land use due to canal construction and water transfer could impact local ecosystems, displacing biodiversity and affecting livelihoods dependent on riverine resources. Available preliminary research highlights the delicate balance between interlinking benefits and risks, emphasising the need for comprehensive impact assessments.

Policymakers must carefully evaluate these projects, prioritising localised water management strategies over large-scale interventions. Integrating advanced hydrological models and environmental safeguards can ensure that interlinking initiatives align with sustainability goals while minimising unintended consequences.

Adaptive Measures and Way Forward

Advanced forecasting tools and data-driven approaches offer hope. Improved weather models, such as those developed under India's National Monsoon Mission, enhance early warning systems, enabling better preparedness regarding water-related risks. Research and development by institutions like the CSIR - Central Salt and Marine Chemicals Research Institute, have introduced cost-effective desalination technologies, providing potable water to coastal and drought-hit regions.

Community engagement through citizen science initiatives, like the Meenachil River and Rain Monitoring Network in Kerala, empowers locals to monitor and manage water resources. Policy interventions must integrate scientific insights with grassroots needs and ecosystem restoration, including reforestation, wetland preservation and sustainable agriculture that can buffer against extreme events.

Startups and research institutions are developing technologies for efficient water use and recycling. Smart irrigation systems, powered by artificial intelligence, are helping farmers optimise water use, while decentralised wastewater treatment plants are addressing urban water woes. These efforts are now scattered, but public-private partnerships are proving instrumental in scaling such solutions. Initiatives under the United Nations' Sustainable Development Goals (SDGs) provide frameworks for achieving water security through global cooperation.

India's water journey from source to sea is at a crossroads. Climate change presents formidable challenges but also opportunities for innovation and resilience. By investing in scientific research and development, community-driven adaptation and sustainable practices, India can safeguard its water heritage for future generations. This collective effort will ensure that water is not merely a resource, but the very lifeline for all.

References

1. Emergency Events Database (EM-DAT) (2025). Centre for Research on the Epidemiology of Disasters.
2. Dhara, C. et al. (2025). Compound weather and climate extremes in the Asian region: Science-informed recommendations for policy. Frontiers in Climate.
3. IPCC (2021). Sixth Assessment Report: Climate Change 2021: The Physical Science Basis.
4. Krishnan, R. et al. (2020). Assessment of climate change over the Indian region: A report of the Ministry of Earth Sciences (MoES), Government of India. Springer Nature.
5. Maurer, J.M. et al. (2019). Acceleration of ice loss across the Himalayas over the past 40 years. Science Advances.
6. Roxy, M.K. et al. (2017). A threefold rise in widespread extreme rain events over central India. Nature Communications.

7. Roxy, M.K. et al. (2018). Impacts of climate change on the Indian summer monsoon. Ministry of Environment, Forest and Climate Change (MoEF&CC), Government of India.
8. Shastri, H. et al. (2015). Impacts of urbanisation on Indian summer monsoon rainfall extremes. Journal of Geophysical Research: Atmospheres.
9. Siddiqui, D. (2015). Water wives. Reuters.
10. UNESCO (2020). The United Nations World Water Development Report 2020: Water and Climate Change.
11. Vaidyanathan, G. (2023). How India is battling deadly rain storms as climate change bites. Nature.
12. Yacob, S. et al. (2025). Dengue Dynamics, Predictions, and Future Increase under Changing Monsoon Climate in India. Scientific Reports.

"To disappear into deep water or to disappear toward a far horizon, to become part of depth of infinity, such is the destiny of man that find its image in the destiny of water."

~Gaston Bachelard

CHAPTER 16

Wetlands Conservation

Somnath Bandyopadhyay

"The water in a vessel is sparkling; the water in the sea is dark.
The small truth has words which are clear;
the great truth has great silence"

~Rabindranath Tagore

Somnath Bandyopadhyay

Dr Somnath Bandyopadhyay is a development ecologist. As an independent researcher based in Vadodara, India, his areas of research include biodiversity conservation, ecosystem services, and sustainable development. He holds a PhD in Environmental Sciences from Jawaharlal Nehru University and a certificate in Environmental Economics from Harvard University.

With over three decades of experience, he has worked with government, civil society, and academia in environmental planning, river rejuvenation studies and rural development programs. He helped establish Nalanda University's School of Ecology and Environment Studies. His talks and writings promote a positive attitude towards the environment, drawing on local poetry, literature, and philosophies.

Currently, he is leading a study to develop a River Health Index for the upper Narmada River, which integrates catchment development and riparian dynamics with the river system.

Wetlands Conservation

Somnath Bandyopadhyay

It was August of 1989. The grades for my MPhil/ PhD course work were good and Jawaharlal Nehru University (JNU) would allow me to opt for a direct attempt at a PhD degree in Environmental Sciences. My supervisor, Prof Brij Gopal[1], however, insisted that I do an MPhil first – he needed to be convinced that I was cut out for the rigours involved in wetland research.

For me, it was literally jumping from the sea to the pond. I already had a Master's in Marine Sciences from Calcutta University, and my dissertation involved extensive field work in the Sundarbans. Moreover, I had spent an additional year as a Junior Research Fellow of the Central Pollution Control Board, collecting and analysing water and sediment samples from the coastal areas of West Bengal and Odisha.

My MPhil dissertation would be on Lake Chilika – a coastal lagoon that was recently designated a Ramsar site. I was also assigned a co-supervisor – Prof Virendra Asthana – who was then involved in mapping coastal areas of India that were likely to submerge due to sea-level rise. Together, they had charted an

1. For a glimpse of the legend, please refer to Bandyopadhyay, Somnath (2022) Looking Back at My Guru "Professor Brij Gopal": An Extraordinary Wetland Ecologist. International Journal of Ecology and Environmental Sciences 48: 3-10.
ISSN: 2320-5199 (Online); https://doi.org/10.55863/ijees.2022.0021

ambitious field exercise for me that involved collecting samples from the benthic zone (i.e. the muddy layer below the water column) of the lake, sieving, processing, and packaging the samples for further analysis in the laboratory.

Individual academic research turned out to be very different from institutional data collection where resources were allocated against a centralised plan, part of which we executed as team members. In this case, the planning, budget and execution were all my responsibility. I was also accountable to two hard taskmasters. Fabricating a core sampler involved discussions with an engineer and an artist, buying materials from hardware stores, finding and engaging an enthusiastic mechanic, and testing the contraption in a local pond. It was bulky and I had to convince a younger cousin to accompany me. However, the biggest challenge was the field itself, which confounded all my previous knowledge and experiences, completely disrupting my expectations, and setting me up for a much deeper engagement with, what we now know as, wetlands.

Nature and Being Human

Lake Chilika is a large coastal depression in Odisha, with water spread over 900 sq. km during the dry season and 1,165 sq. km during the wet season. At least 50 small rivulets drain into this lagoon, along with a few rivers such as the Daya and the Bhargavi. This brackishwater lagoon is separated from the Bay of Bengal by a sand bar that had a narrow opening near Arakhakuda back then. The complex interplay of fresh and sea water created an invisible mosaic of salinity that changed like a kaleidoscope over the seasons. The benthic fauna that I set out to sample held a promise of revealing a part of this interesting pattern. Maybe even explain it a bit. It was wetland science, exactly as I was meant to study.

The bumpy ride in a state transport bus from Puri to Satpada took about 4 hours to complete 70 km. We were dropped off near the coast, some distance before Satpada. After walking through a km of desolate, sandy landscape, we suddenly came to a small fishing hamlet swarming with children – their fathers were out fishing, and their mothers busy at home. We realised that we had left civilisation behind and basic necessities would be our first challenge. Luckily, we spotted a tin-shed, amidst the clutter of mud huts, with a weathervane swivelling on top.

A young officer of the weather department was more than happy to see us. Although we spoke different languages, we had a civilisational connect. After offering us some food, he introduced us to an adolescent fisherman who agreed to take us out on his boat. We decided to sample the long narrow channel that connected the Lake with the sea. The bed being sandy, the sampler would not work. But the water was not too deep either, although there was a perceptible current. I decided to jump into the water, scoop up the sediment and put it in the sieves held out by my cousin, who stayed on the boat. My first field samples were being collected.

As I surfaced with another grab of sediments, I was startled to hear the shrieks and see the vigorous gesticulations of the boatman and my cousin. Instinctively, I dropped the sample, grabbed the edge of the boat and hauled myself into it, just in time to see a large snake slithering along the boat making a splash with its flat tail at the spot where I had been standing a moment ago. All three of us froze as we absorbed the import of that close brush.

It was difficult to enter the water immediately thereafter and we decided to row further towards the sea-mouth, the *muggermukh*, where the current was anyway dragging us. The conversation was predictably around large predators in the ecosystem. Suddenly we saw something floating alongside. The boatman picked it up to reveal the remains of a fish. I thought it was the work of a predator and conjectured what may have caused it. The boatman gave me a demonstration. He pulled out a fish from below a cloth cover, sliced it into two with a heavy blade and dragged the bleeding end in the water. Within moments, the dark outlines of a large fish emerged from nowhere and began following the bloody trail. In one swift motion, the boatman pulled out a short fishing spear, took aim and threw it pinning the fish to its death.

As he hauled the large *bhetki* (*Lates calcarifer*), I inquired if he would sell it. "Yes, at ₹5,000", he said.

"I only have ₹50", I replied, quite dejected.

"Okay, that will do", he said, to my utter surprise. Clearly, he didn't quite know the difference. He also promised to cook me a complimentary dinner on the sand bar, where we were now stranded till the tide turned. The breeze was already cool though the sun was yet to set. Up on the sandbar, I watched the *bhetki* as it entered the flame created with twigs and coal in a makeshift oven on which a pan of rice was put to boil simultaneously. I could hear my cousin talking animatedly with a bunch of male kids – less than 10 years of age, not wearing any clothes – splashing and giggling at the water's edge.

My cousin's ecstatic shouts startled me out of my thoughts as I looked towards him, and the waters in the channel beyond. Something had emerged a moment ago, and had now dropped back into the waters leaving a splash and ripples. I ran down to his side, just in time to see two Irrawaddy dolphins (*Orcaella brevirostris*) make a rare full breach. A split-second of natural romantic ballet, illuminated by the golden rays of the setting sun, framed against the dark bushes on the far shore, the blue waters of the channel below and white sands in the foreground, remain etched in my mind's album. The ballet continued for a while, with a troupe of about 8-10 dolphins, their chirps and whistles mingling with the joyous cries of children playing alongside, each oblivious of the presence of the other. We slumped on the beach, absorbing the images of a lifetime as darkness settled around us.

The heavens were now lit with millions of stars and the ocean surface was glistening in reflected glory. The fire crackled and glowing embers rose to kiss the stars. Food was almost ready when I noticed a few smaller fishes were also put to boil. These were gifted by the children who were busy catching fish even while they played in the water. The piece of cloth they were seemingly playing with in the water served to sieve out a few fish from the channel.

"They are collecting gifts for their brides", my cousin explained. "A mass wedding has been arranged for these kids tomorrow morning". Individual wedding celebrations would be far too expensive. Many of the children would likely not live much beyond their adolescence or youth. As a closed community, it was just a minor shuffling of the way in which they played, slept, hunted and died.

Nature and Civic Amenities

Back in Kolkata, then called Calcutta, I spent time with Dr Dhrubajyoti Ghosh, a senior official at the Department of Environment, Government of West Bengal. Together, we travelled to different parts of what is now known as the East Kolkata Wetlands (EKW).

Many years ago, as a civil engineer, Dhrubajyoti was given the task to figure out how Kolkata, and indeed the Hooghly River, survived without any significant sanitation infrastructure. Kolkata was growing fast, and, in comparison, the sewerage infrastructure was increasingly becoming dysfunctional and irrelevant. Yet, unlike 19th century London, there were no major cholera, or any other public health epidemic reported. He was surprised to find that the city sewage was indeed flowing mostly in open drains, but instead of the River Hooghly it was primarily flowing in the opposite direction into a large landlocked estuary of the River Bidyadhari, known as the Salt Lakes.

The remnants of many Gangetic distributaries between the Hooghly and the Ganga manifested as a mosaic of marshy lands interspersed with open waterbodies, locally known as '*beels*'. These wetlands provided a gradient that sustained the local population with abundant harvest of rice, reeds and fish, shared with birds and aquatic animals. Closer to Kolkata, though, the *beels* interacted with manmade canals – *khals* – to produce an even more complex landscape, almost the size of the city itself, to naturally purify its liquid waste. Dhrubajyoti named these the East Kolkata Wetlands (EKW) and designated their function as 'the kidneys of the city.'

Now specialising in ecological systems, Dhrubajyoti demonstrated the importance of urban wetlands not only for biodiversity but also for wastewater treatment, food security, and sustainable livelihoods. This unique wastewater-fed aquaculture system treats 78% of Kolkata's sewage naturally, saving the city approximately ₹4,700 million annually in treatment costs. His research revealed that these wetlands treated over 1,000 million litres of sewage daily and recovered resources through ecological processes, thereby supporting one of the most densely populated areas in the world. He recognised the role of sunlight, endemic biodiversity and ecological processes in subsidising a growing economy and sustaining the aspirations of millions who migrate to eastern India's only urban centre.

He also admired the knowledge of the poor people whose lives were connected to this process. The East Kolkata Wetlands (EKW) support the livelihoods of an estimated 150,000 to 200,000 people through a unique blend of traditional aquaculture and agriculture. With around 245–260 shallow fishponds, the wetlands produce nearly 20,000 metric tonnes of fish annually, valued at over ₹33,000 per hectare per year. In addition, vegetables and paddy are cultivated using urban waste and wastewater, with daily vegetable production reaching 150 metric tonnes and paddy fields delivering handsome returns—up to ₹148,748 per hectare annually. The total annual economic value of the wetlands is estimated at over ₹924 million across 10,134 hectares.

Dhrubajyoti was amazed by the efficiency of Nature's waste recycling, the effortless circular economy in-built within Nature, and its stupendous impact on public health and nutrition. His faith in Nature, drawn from a deep understanding of its functions, far outstripped his training in environmental sanitation. I, on the other hand, was a novice environmental scientist trained to study Nature *sans* any human interference. A utopia that not only did not exist but was absurd in its framing.

Considering Wetlands as Wastelands

The East Kolkata Wetlands (EKW), unfortunately, were just on the eastern fringes of the city of Kolkata, and stood in the way of its expansion, particularly because its western boundaries were marked by the mighty Hooghly River. When I met Dhrubajyoti in September 1989, the battle for the EKW had become even more unequal with the powerful real estate lobby moving in to reclaim areas for urban expansion. Salt Lake City was a well-established affluent township by the 1980s, and the Rajarhat New Town would soon be sanctioned in the 1990s. All at the expense of the EKW.

Draining and reclamation of wetlands is not new. In the early 20th century, wetlands were widely viewed as unproductive wastelands—breeding grounds for disease and obstacles to agriculture and development. This negative perception led to massive drainage campaigns, especially across Europe, North America and Asia, aimed at converting wetlands into farmland.

Across Europe, wetlands have been drastically reshaped by centuries of reclamation. In the Netherlands, wetland drainage began with the creation of polders like the Beemster in 1612, where windmills pumped out lakes to create farmland. Soon thereafter, the Fens of East Anglia in England were drained; while France's Marais Poitevin was slowly drained and turned into farmland, crisscrossed by canals. Meanwhile, in countries like Ireland, Scotland and Finland, bogs were heavily mined for peat fuel in the 18th and 19th centuries, leaving lasting scars on the landscape.

In the U.S., laws passed in 1849, 1850, and 1860 gave states control over swamplands to drain them for farming. This led to the loss of huge wetland areas like the Mississippi Delta, Indiana's Kankakee Marsh and parts of the Florida Everglades. By 1900, over half of the country's wetlands had been drained for farming and for cities. Major rivers like the Illinois and areas like the Atchafalaya Basin were also reshaped. The Everglades were long seen as dangerous and unhealthy, and serious drainage efforts began in the 1930s, using canals and levees to control the water. It was all part of the idea that the wild lands of the New World had to be 'tamed'.

Later, urban expansion, infrastructure projects and industrialisation began to take an even greater toll. Wetlands were filled to make way for cities, highways and factories, especially in growing economies of Asia and South America. In the Global

South, aquaculture—particularly shrimp farming—became a major driver of mangrove destruction. At the same time, the construction of dams and irrigation systems altered natural water flows, starving wetlands of the seasonal floods they depended on. Add to this the effects of pollution, invasive species and now climate change, and it's easy to see how so much has been lost.

Recent studies reveal that we've lost an estimated 64% of the world's wetlands since 1900[1], and the story of their disappearance is one of relentless human pressure, poor perception, and shifting priorities. Worse, if we look back at 1700, we may have lost as much 87%. Perhaps the most sobering part the ongoing is loss — wetlands are disappearing three times faster than forests today.

What Exactly Are We Losing?

Losing wetlands means access to fresh water is eroding for 1-2 billion people worldwide, while flood control, carbon storage and traditional wetland livelihoods all suffer. Biodiversity has also been affected, as populations of freshwater species have declined by 76% between 1970 and 2010 according to WWF's Living Planet Index.

Wetlands support approximately 40% of all plant and animal species, including many endangered ones, while only covering about 6% of the land surface. This remarkable concentration of biodiversity highlights wetlands as ecological powerhouses, where the interplay of land and water creates unique niches that sustain species ranging from microscopic organisms to large mammals.

Lake Chilika, a Ramsar site since 1982, faced issues like reduced salinity, siltation and declining fish species, due to the

1. Davidson, Nick. (2014). How much wetland has the world lost? Long-term and recent trends in global wetland area. Marine and Freshwater Research. 65. 934-941. 10.1071/MF14173.

closure of its mouth. In 1993, it was added to the Montreux Record for experiencing serious ecological decline, including blocked sea-water exchange, dominance of freshwater inflow, and rapid spread of invasive weeds. This caused a drastic drop in salinity, collapse of fisheries and a significant decline in bird populations, with migratory birds decreasing from over 1 million in the 1980s to under 200,000 by the early 1990s. Nalabana Island also lost most of its avian visitors.

Beyond their biological functions, wetlands play a broader role in maintaining environmental balance. They act as natural buffers against extreme weather events, absorbing excess floodwaters and releasing them gradually to mitigate flood impacts and maintain stream flows during droughts. Additionally, wetlands contribute to groundwater recharge and erosion control, enhancing the resilience of landscapes to storms and rising sea levels. Wetlands act as natural water purifiers, filtering sediment and absorbing many pollutants in surface waters. They also serve as carbon sinks, storing significant amounts of carbon in their vegetation and soils, which is crucial for mitigating climate change.

In recent years, economists have started to put a price tag on what wetlands do for us — helping people understand what we lose when wetlands are destroyed, and how costly it can be to restore them later. Wetlands are incredibly valuable. In 1997, their global services were estimated at $14.9 trillion a year[1]. A revised estimate suggests that wetlands contribute over $47 trillion annually in ecosystem services[2]. Wetlands cover just 6%

---

1. Costanza, R., d'Arge, R., de Groot, R. *et al.* (1997). The value of the world's ecosystem services and natural capital. *Nature* 387, 253–260. DOI: https://doi.org/10.1038/387253a0
2. Costanza, R., de Groot, R., Sutton, P., van der Ploeg, S., Anderson, S. J., Kubiszewski, I., Farber, S., & Turner, R. K. (2014). Changes in the global value of ecosystem services. *Global Environmental Change*, 26(1), 152–158. DOI: 10.1016/j.gloenvcha.2014.04.002.

of Earth's surface, but account for more than 40% of the total value of all ecosystem services worldwide—making them some of the most important and undervalued ecosystems we have.

Struggles to Protect Wetlands

Wetlands have been integral to human societies for centuries, providing not only resources but also spiritual and inspirational values. They are often tied to traditional practices, community livelihoods and indigenous knowledge systems. Resistance from local communities, who relied on the wetlands for fishing, fowling and grazing, was often fierce. In 17th century England, locals fiercely resisted the drainage of East Anglia wetlands led by Dutch engineer Cornelius Vermuyden, hired by James I and Charles I. 'Fenlanders' sabotaged dikes to protect their livelihoods of fishing, fowling and peat-cutting.

Today, the East Kolkata Wetlands (EKW) represent one of India's most enduring and iconic environmental struggles — a continuous fight to protect a living ecosystem from the relentless pressures of urban expansion. Since the early 1990s, powerful real estate interests — often operating with tacit state support — have sought to carve this land for trade centres, luxury housing and infrastructure projects.

At the heart of the resistance stood Dr. Dhrubajyoti Ghosh, whose pioneering work revealed how the wetlands transformed urban sewage into fish and crops using natural ecological processes. Dhrubajyoti's research not only gave the wetlands scientific credibility, but also galvanised public awareness, laying the foundation for legal and institutional protection.

The struggle gained momentum in 1992, when the NGO PUBLIC (People United for Better Living in Calcutta) filed a Public Interest Litigation (PIL) to block the allocation of 227 acres of wetland for a proposed World Trade Centre. In a landmark ruling, Justice Umesh Chandra Banerjee of the Calcutta High Court recognised the wetlands' vital ecological and economic functions, prohibiting further development without proper assessment. This judgment formally established the EKW's legal identity, setting a powerful precedent — though its enforcement has remained inconsistent.

In a major victory for Dhrubajyoti, the EKW was designated a Ramsar Site on August 19, 2002, earning international recognition and placing India under global obligations to protect it. This was followed by the East Kolkata Wetlands (Management and Conservation) Act, 2006, which established the East Kolkata Wetlands Management Authority (EKWMA). These milestones marked crucial gains: they legally bound the state to protect the wetlands, formalised governance mechanisms and spotlighted the site's global ecological significance.

Yet, the struggle continues. Despite its Ramsar status, the EKW has lost nearly 36% of its area, shrinking from 12,500 hectares to around 8,000 hectares due to illegal encroachments, industrial dumping and stalled enforcement. Budgetary allocations remain negligible, while proposals like solid waste dumping and new flyovers continue to threaten the ecosystem. The battle for EKW is far from over, but its history is proof that scientific vision, legal action and grassroots resistance can together hold the line against concrete ambitions.

The Lake Chilika was also degrading rapidly, despite its Ramsar status. However, two important steps were taken. The Chilika Development Authority was established in 1991, and secondly, the degradation of the wetland was acknowledged by placing the site under the Montreux Record in 1993. By 2000, a breakthrough was achieved by opening a new mouth to the sea near Satpada, restoring tidal flow and salinity, reviving the lake's productivity and restoring Nalabana as a vital hub along the Central Asian Flyway. The impacts of restoration were dramatic. Fish and prawn yields increased more than threefold, benefitting over 150,000 traditional fisherfolk who depend on the lake. The lagoon also saw a revival in biodiversity, again becoming a haven for over 1 million migratory birds each winter. The return of the Irrawaddy dolphin, the sightings of which had become quite rare during the 1990s, marked another ecological milestone. In 2002, Chilika became the first wetland in the world to be removed from the Montreux Record, marking a globally celebrated restoration success driven by scientific intervention, hydrological engineering and community action.

Mangalajodi, a quiet village on the northern edge of Chilika Lake, is perhaps the best example of community action. Today, it hosts over 300,000 birds each winter and shelters more than 200 species, including spectacular migrants like black-tailed godwits and pintail ducks. But just three decades ago, this was a hunting ground where birds were relentlessly trapped and killed for food and profit. Among the hunters was Nandakishore Bhujbal, who was haunted by the memory of shooting an egret while it was in the process of building its nest; a moment that changed him forever. Facing ridicule and resistance, he pleaded with fellow hunters to stop, until his guilt and passion touched even hardened hunters like Madhu Behera, who ultimately laid down their weapons. The turning point came in 1996, when the NGO Wild Orissa began sensitising the community, leading to the formation of the Sri Sri Mahavir Pakshi Suraksha Samiti by 2000. Reinventing themselves as bird guides, patrollers and stewards, these former hunters transformed Mangalajodi into a model of grassroots conservation and eco-tourism.

Ramsar, Wetland Conservation and Birds

The Ramsar Convention, formally known as the Convention on Wetlands of International Importance especially as Waterfowl Habitat, was adopted on February 2, 1971, in Ramsar, Iran. The treaty came into force on December 21, 1975, and has since become a cornerstone of international efforts to conserve wetlands. Under the Ramsar Convention 172 countries (Contracting Parties) have designated 2,531 Ramsar sites (in February 2025) covering over 2.6 million square kilometres.

The historical context leading to the Ramsar Convention involved a series of events and efforts by key personalities such as Dr Hans Lukas (Luc) Hoffmann (1923-2016) and Sir Peter Scott (1909-1989), a Swiss and a British ornithologist respectively. Hoffmann's proposal to the International Union for Conservation of Nature (IUCN) led to the launch of Project MAR in 1960, which aimed to conserve and manage marshes, bogs and other wetlands. This project involved organisations like the International Council for Bird Preservation (ICBP, now BirdLife International) and the International Wildfowl Research Bureau (IWRB, now Wetlands International).

Luc Hoffmann, heir to the Hoffmann-La Roche pharmaceutical legacy, chose to channel his life's passion not into business, but into birds. From a young age, he was captivated by ornithology, and this fascination shaped his life's work. He was raised in a household rooted in art and philanthropy with his mother being a sculptor and his father an art collector. Hoffmann charted a different course: the conservation of birds and their habitats. His scientific dedication led him to establish the Tour du Valat research station in the Camargue wetlands of southern France in 1954. This became a pioneering centre for wetland ecology long before the field gained global traction. He carried out detailed studies of flamingos and wetland birds, linking rigorous ornithological research with wetland management.

Sir Peter Scott was the son of Antarctic explorer Robert Falcon Scott and sculptor Kathleen Bruce. Scott founded the Severn Wildfowl Trust (now the Wildfowl & Wetlands Trust) in 1946, which became a key site for saving species like the Hawaiian goose from extinction. He was a co-founder of the World Wildlife Fund (WWF) and designed its iconic panda logo. His legacy includes the establishment of the IUCN Red List of Threatened Species.

Birds have been at the heart of wetland conservation since the very beginning of the Ramsar Convention, which was formally titled the *Convention on Wetlands of International Importance Especially as Waterfowl Habitat* when signed in 1971. This reflects the founding vision of wetlands as critical habitats for migratory waterbirds. Wetlands across the globe serve as essential breeding, nesting, feeding and resting grounds for birds, particularly along key migratory flyways like the East Asian–Australasian and Central Asian Flyways, which support tens of millions of birds each year. In fact, more than one-third of North American bird species use wetlands; in the U.S., around 138 species are considered wetland-dependent. Birds not only benefit from wetlands, they also act as indicators of ecosystem health, with alarming declines in migratory waterbird populations mirroring the global loss of 35% of wetlands since 1970.

Birds also remain central to how wetlands are formally recognised and protected under international frameworks. Criterion 5 of the Ramsar Convention's nine criteria for identifying wetlands of international importance, explicitly states that *a wetland should be considered internationally important if it regularly supports 20,000 or more waterbirds.* This criterion has led to the designation of key sites across the world, where such congregations occur regularly, ensuring legal recognition and conservation attention. Even as Ramsar has evolved to include broader ecological and human values, birds continue to play a vital role as flagship species, as tools for scientific monitoring and as enduring symbols of the deep connection between wetlands and life.

Square Peg in a Round Hole?

The modern environmental movement is rooted in guilt and remorse. Aldo Leopold built a compelling ethical narrative of sharing the earth with other species goaded by the tormenting vision of the eyes of a dying mother wolf[1]. Rachel Carson invokes the ubiquitous and much loved robin to drive home the cruelty of industrial chemicals on their reproductive capabilities. Luc Hoffman grew up with Calvinist values, choosing a strong work ethic, humility and a life lived for the glory of God.

The remorse of bird hunters in Mangalajodi is remarkable only because they connect with their civilisational roots established thousands of years ago. The sarus crane (*Grus antigone*) is revered as a symbol of unconditional love and marital fidelity, believed to mate for life and pine the loss of its mate to the point of starvation. It features prominently in Indian mythology, with stories like the legend of Valmiki, who was inspired to write

1. Aldo Leopold underwent a pivotal shift after shooting a wolf in New Mexico. Witnessing the "fierce green fire dying in her eyes" profoundly affected him, leading him to question the role of predators in ecosystems and ultimately develop his "land ethic" philosophy. This transformation not only reshaped his views on hunting but also influenced his advocacy for sustainable wildlife management and wilderness preservation. Leopold's writings, particularly in *A Sand County Almanac*, resonated with both scientists and the general public, including elites who began to see nature as an interconnected system rather than a resource to be exploited.

the Ramayana after witnessing a grieving sarus crane[1]. The bird is also mentioned in the works of ancient poets like Kalidasa and is considered sacred by the Gond tribal people.

Restoring Lake Chilika required scientific understanding of the relevant issues, political resolve to address the fundamental causes, as well as a robust institutional mechanism for implementation. Conversely, the East Kolkata Wetlands continue to face challenges despite having basic legal and institutional frameworks in place. Acknowledging the role of abundant sunlight and natural algae raises questions about the necessity of sewage treatment plants (STPs). Similarly, recognizing the knowledge of indigenous communities might spark debates about modern technology, which is integral to contemporary economies. Furthermore, the mere existence of these wetlands could hinder real estate development opportunities. Future challenges for Lake Chilika may include aquaculture, tourism infrastructure and even 'well-meaning' restoration projects.

India's engagement with the Ramsar Convention began on February 1, 1982, with the designation of Chilika Lake in Odisha and Keoladeo National Park in Rajasthan as its first Ramsar sites. From 1982 to 2013, India added 26 sites to the list, reflecting a steady but modest pace of wetland conservation. However, between 2014 and 2022, there was a significant acceleration,

1. In his emotional state, the sage Valmiki spontaneously uttered a curse against the *nishada* (hunter):
"मा निषाद प्रतिष्ठां त्वमगमः शाश्वतीः समाः।
यत्क्रौञ्चमिथुनादेकमवधीः काममोहितम्॥"
Transliterated as:
"mā niṣāda pratiṣṭhāṁ tvamagamaḥ śāśvatīḥ samāḥ
yat krauñcamithunādekam avadhīḥ kāmamohitam"
This curse roughly translates to:
"You will find no rest for the long years of Eternity, for you killed a bird in love and unsuspecting".
Composed in perfect meter, it became the first *shloka* in Sanskrit literature, marking the birth of classical poetry. The sarus crane thus became an enduring symbol of love, loss, and devotion, while Valmiki came to be revered as Adi Kavi—the first poet.

with 49 new wetlands designated as Ramsar sites. By 2022, India had reached a milestone of 75 Ramsar sites covering over 13 lakh hectares. This rapid increase continued into 2024 and early 2025, with additional designations bringing the total to 89 Ramsar sites.

At least 42 among these are also legally protected as wildlife sanctuaries, bird sanctuaries, conservation reserves or national parks. Examples include Keoladeo National Park, Rajasthan; Kolleru Lake Wildlife Sanctuary, Andhra Pradesh; and Chilika Lake, which includes Nalabana Bird Sanctuary, among others. The recent additions like Karaivetti Bird Sanctuary and Nanjarayan Bird Sanctuary in Tamil Nadu continue this trend of overlap between Ramsar designation and protected area (PA) status.

A Ramsar designation recognises international ecological importance under the Ramsar Convention, focusing on wise use, wetland functionality, and biodiversity, especially for migratory waterbirds. However, Ramsar status alone does not automatically confer legal protection under Indian environmental law. It mainly creates a moral and international obligation. When a wetland is also declared a protected area under Indian laws, like the Wildlife Protection Act, 1972, it gains stronger legal backing, including restrictions on land use, development, hunting and access. PAs have defined management plans, dedicated enforcement mechanisms and legal penalties for violations. Customary or traditional use, such as those of the fishermen in the Chilika mouth region, become illegal.

Hoffmann was instrumental in demonstrating how wetlands were vital not just for biodiversity, but for sustainable human development. Dhrubajyoti spent his life working tirelessly to save the East Kolkata Wetlands and the ecological knowledge it held. The East Kolkata Wetlands were declared a Ramsar site in 2002, an international instrument created by Hoffman more than three decades earlier.

Hoffman passed away in 2016, and the same year, Dhrubajyoti was honoured with the Luc Hoffmann Award for Excellence in Ecosystem Management, granted by the International Union for Conservation of Nature (IUCN) to recognise his exceptional initiative in ecosystem management. Two years later, Dr Dhrubajyoti Ghosh himself passed away.

It was sheer providence that I was with Dr Dhrubajyoti Ghosh when he received the news of receiving the Luc Hoffman award. It was an autumn evening in 2016 at the make-shift campus of the project to revive the Nalanda University. Earlier that day, he spoke to the students at the Rajgir International Convention Centre, cautioning them against the tendency to search where there is light, referring to an old parable of Molla Nasiruddin[1]. He stood on the empty plot of land that would be the rejuvenated campus one day, and declared 'this is more important than anything we have built so far.' He opened his email before dinner and saw the message of his award. He was happy that the world was taking note of the dirty job being done by Nature, not just applaud or grieve the iconic wildlife, mostly birds.

The award had a broader significance. Hoffman began with bird conservation and expanded to understanding the wider ecological functions of wetlands. Dhrubajyoti started with the ecological functions of wetlands and came to appreciate the

1. One evening, Molla Nasiruddin was on his hands and knees under a streetlight, searching for something. A passerby noticed him and asked, "Molla, what are you looking for?"
Nasiruddin replied, "I'm searching for the key to my house."
The passerby, wanting to help, joined him in the search. After a while, the man asked, "Where exactly did you drop the key?"
Nasiruddin pointed to a dark area away from the streetlight and said, "Over there, inside my house."
The man, puzzled, exclaimed, "Then why are you looking for it here?"
Nasiruddin calmly replied, "Because there's more light here."
This humorous tale carries a deeper meaning. It reminds us that we often look for solutions or answers in places that are convenient or comfortable, rather than where the real problem lies. It's a gentle nudge to confront challenges directly, even if it means stepping into the metaphorical darkness.

ecological knowledge of its traditional users. Together, they leave a significant message for the next generation.

Wetland conservation is more than protection of birds, and also of people living there. Wetland conservation is about wisdom; not only in how we cherish and restore these living waters, but in how we listen to their silence. Above all, wetland conservation calls for determination, dedication and resolve, in working in harmony with a shared sense of purpose.

CHAPTER 17

Need for Sustainability in the Ocean Realm

Purnima Jalihal

"The heart of man is very much like the sea, it has its storms, it has its tides and in its depths it has its pearls too."

~Vincent van Gogh

Purnima Jalihal

Dr Purnima Jalihal is a senior scientist who heads the Energy and Fresh Water group in the National Institute of Ocean Technology, India. She has led device developments for harnessing ocean energy from waves, marine currents and ocean thermal gradient (OTEC) and has played a major role in developing the ocean thermal desalination technology.

She has a PhD in Civil Engineering from Duke University, USA. She is on the committees of several Indian Government organisations, related to water and renewable energy. She is currently the Vice Chair of the Executive Committee of the Ocean Energy Systems (OES) Technology Collaboration Program (TCP) under the International Energy Agency (IEA).

She was awarded the Vishwakarma Medal in 2006 by the Indian National Science Academy for her contribution to cold water pipeline of the first LTTD plant in Kavaratti in Lakshadweep. She also received the Uehara Prize 2019 given by the International OTEC community for Contribution to the Development of OTEC. She was designated the EU led Clean Energy Mission Innovation Champion for India, 2020.

Need for Sustainability in the Ocean Realm

Purnima Jalihal

White froth breaks on beaches, seemingly incessantly, with great intensity. Far away, the horizon appears to be demarcating the boundaries to this vast expanse of water. The oceans remain a mystery in many ways. They pervade all our known senses and sensibilities.

They have long been in the discourse of literature, music and dance in all cultures. They are a part of scientific study in biology, chemistry, physics and geology. There is a vast area called oceanography devoted to the behaviour of oceans in the form of tides, waves, currents, temperatures.

In India, the ancient *samudra manthan* or churning of the oceans for nectar, is a well known part of mythology. Metaphorically, the churning reveals a storehouse of wealth. We know that the Earth is made up of 70% of water, which is primarily in the oceans. Apart from the knowledge of what surrounds us, mankind in its constant quest, looks for resources and ways to harness this wealth sustainably from the ocean in order to augment that from the land. However, technologies for the ocean are far more complex than those employed on land. Harnessing resources from the oceans is very challenging.

Chemical and Physical Oceanography

The six most abundant ions in sea water are chloride, sodium, sulphate, magnesium, calcium and potassium. Inorganic elements like carbon and boron are other major dissolved substances. Many dissolved chemical constituents are responsible for the growth of marine organisms. Physical mixing in the ocean tends to homogenise the chemical composition of sea water. The delicate balance of the ecosystems of the oceans is very evident, and a huge number of factors affect every aspect of the oceans.

Oceanic dimensions range from around 1,500 km width to more than 13,000 km. Typical depth can be 3-4 km, with the features of the sea floor varying considerably. Sound is the only signal that can be used for remote sensing of the ocean below a depth of few tens of meters. It can be used to measure properties of the sea floor, depth, temperatures and currents.

Ocean Resources

Oil and Gas

Gas and oil form in the seabed over several million years. Remains of animals and plants sink to the ocean floor. Combined with other particles they get buried and compressed to form sediments which can be several kilometres thick on the ocean floor. The Earth's pressure and temperature aids in converting this biomass into hydrocarbons. Depending on the ambient condition, oil or natural gas gets formed. Today's fossil fuels are 15-600 million years old. Industrial oil extraction began in mid-19th century. Consumption of natural gas is also steadily growing. Liquefied natural gas (LNG) plays a major role in humankind's interest in the sea. Cooled LNG is shipped to the shores and is a major global trade component. Today offshore oil prospecting is well established but costs remain high due to offshore and deep-water operations.

As the world's reserves of oil and gas run short and prices increase, currently unprofitable sources will become more economic. Nobody knows for certain how long global reserves and resources of oil and gas will last – particularly because it is difficult to predict future consumption trends. From today's perspective, resources of natural gas will probably be adequate to ensure supply into the second half of this century. But if natural gas is used to power motor vehicles or generate electricity in power stations, the reserves could be exhausted quicker.

While large economies are running on offshore oil and gas, these are fossil fuels and have two distinct disadvantages: they damage the environment and they get depleted.

Marine Biodiversity

What are the other resources that are harnessed? Food, energy, water and minerals have been amply available in the seas and these fall into the blue economy paradigm. Marine biodiversity is vast and yet to be explored. With the long Indian coastline and over 30% of the population living on the coast, sea food is a primary resource. Efforts to monetise and scale up operations for harnessing sea food continue. Aquaculture farms and technological improvements in shallow and coastal waters have been made. However, our knowledge of deep-sea biodiversity is still very scant and the efforts to get more data are challenging. Seaweeds and corals are commonly known, but organisms like spiral whip coral colonies have been found even in deep water and are yet to be understood fully.

Energy

While oil and gas are fossil fuels, can the ocean be a storehouse of other forms of energy?

The incessant waves are caused by winds blowing on the surface. Due to the irregular nature of wind, waves are also an irregular phenomenon and their intensity varies with seasons and location.

Waves have fired the imagination of many a mariner because their energy can be felt physically as we sail across the seas. Waves can be a source of energy and can be harnessed. Ocean currents are driven by wind and solar heating of the water near the equator, although some currents result from variation in water density and salinity. Ocean currents are much slower than winds. Water is about 835 times denser than wind, so a small water flow can be equivalent to wind with high velocity. These marine currents are also possible sources of energy.

However, waves and currents change with season both in magnitude and direction. The water depth and sea bed profile play a role in the energy they carry. When we harness energy from ocean currents, we are actually tapping the kinetic energy of the moving water. Additionally, potential energy can also be tapped from the difference in height (or head) between low and high tides.

Tides are caused due to gravitational effects. Tidal flow occurs due to the moving mass of water with speed and direction determined by the gravitational forces of the sun and moon and the centrifugal and inertial forces of the Earth's waters. Tides move large quantities of water twice each day, and thus can be utilised for harnessing power.

Countries like India have low wind and wave climates. However, being close to the equator and temperatures being high, the sea surface is always fairly warm. The ocean's temperature varies with depth. The temperature at 1,000 metre water depth could be as low as 6°C - 8°C. This difference in temperature between the sea surface and a deeper depth can be utilised to harness energy. This is called Ocean Thermal Energy Conversion or OTEC.

The ocean is vast and completely unobstructed for harnessing wind. While wind turbines on land have become common, mounting them offshore poses technical and commercial challenges. However, offshore wind is being seen as a possible resource in the future.

Water

While 70% of the earth's surface is covered by sea water, fresh water is merely 2%. With ever-increasing water stress, we are turning to the ocean to convert saline water to fresh potable water. Various methods of desalination of seawater are possible and have the potential to reap this plentiful resource. The water of the oceans is the single most important factor in the existence of the hydrological cycle. The evapotranspiration, the connectivity to the land mass through clouds and rain and the overall environmental balance all have seawater as the core important element.

Minerals and Gas Hydrates

The seabed, especially in some deeper regions like the Central Indian Ocean basin, hosts a huge variety of minerals like cobalt, manganese, nickel and iron in the form of polymetallic nodules. Apart from wealth generation, they are also strategically important from the perspective of areas allocated to different countries by the International Seabed Authority. Gas hydrates have been identified in many parts of the world including the KG basin in India. These are crystalline solids formed of water and gas and contain huge amounts of methane. These are considered to be a major energy resource, although difficult to harness.

Additionally marine microalgae have been shown to be efficient in utilising carbon dioxide and have shown potential to yield higher oil production per hectare, compared to land crops. This shows promise as a new source of renewable energy.

Current Issues

It is clear that the oceans are a storehouse of resources, with their own challenges and with innumerable hidden mysteries. It is unfortunate that this very vast beautiful entity of our Earth is in distress. What are the issues which plague the marine scenario today? It is now undisputably clear that the global climate is changing. But what is its impact on the oceans?

The three primary fallouts are rise in global temperature, change in precipitation patterns and rise in sea levels. These three prominent signals of climate change convert to regional scale hydrological changes like modification in water availability, water demand, hydrological extremes of floods and droughts, deterioration in water quality, salinity intrusion in coastal aquifers, groundwater recharge and other related phenomena [1]. It is evident that any atmospheric temperature increase or change in precipitation is going to affect river run off, and eventually the various facets of the hydrological cycle in which the oceans play a large role. While the threat of climate change affects the entire globe, India's vulnerability to climate change impacts is particularly high, since a large population depends on rain-fed agriculture, and at least 30% of the population lives along the coastline. They are at risk due to sea level rise and extreme weather events.

Shoreline Changes

The vulnerability of coastal areas to sea level rise is compounded by anthropogenic activities leading to coastal erosion, posing a major threat to ecologically important areas as well as human habitats. Among the coastal resources, sandy beaches are highly susceptible to short term changes due to storm surges. This in turn affects turtle nesting sites and other unique intertidal fauna dependent on these beaches [2]. The increased frequency of storms and sea-level rise due to climate change could result in beach loss, which is also an important habitat for turtle nesting.

In this context, sustainable shore protection measures are now important in areas experiencing severe erosion. Often, anthropogenic intervention for erosion mitigation is in fact detrimental to adjoining areas and thus coastal vulnerability is a primary factor to consider for any human or environmental intervention. Apart from shoreline changes, there are other anthropogenic causes which affect and pollute the oceans.

Shipping and Effluent Pollution

Globally, approximately 2,950 million tankers of crude oil and oil products are transported annually by sea. Occasionally some oil spills occur and cause marine pollution worldwide. Operational or accidental release of oil into the sea has caused harmful effects on marine environment and marine life [3]. Petroleum hydrocarbons are mostly toxic, mutagenic and carcinogenic and cause serious environmental problems due to their persistence in the environment and bioaccumulation.

Industries release large quantities of effluents and unfortunately it is easy to dispose effluents in water bodies like rivers or the ocean. Municipal and household waste is harmful due to coli form and faecal matter and other microbiological contaminants. However industrial waste can contain a mixture of different chemical and toxic substances. When heavy metals like mercury, lead, cadmium etc. are present, they have no role in biochemical function and hence can be very harmful to flora and fauna. Treatment systems need to be really effective. Constant discharge of large quanta of such effluents is having detrimental impacts on coastal waters. Industries are actually using 12-15% of the total available water and in turn generate and release highly polluted water. As against domestic waste which is largely organic, industrial waste can contain different components which vary from industry to industry. It is now being mandated to incorporate zero liquid discharge (ZLD) mechanisms, but treatment systems need to be site specific and hence cannot be generalised. Human activities have influenced composition of oceans, both locally and globally. Discharge of untreated sewage or fertilisers in coastal waters leads to increased phytoplankton growth, high levels of dissolved and particulate organic materials, decreased light penetration, altering the community structures of marine organisms.

Ocean acidification is caused by carbon dioxide emissions. Oceans absorb substantial amounts of carbon dioxide and thereby consume a large portion of greenhouse gases which are released by human activity and due to this, the oceans are becoming acidic. The consequences of this acidification on marine life are being studied and need long term monitoring. Oxygen dead zones are being created making it dangerous for living organisms under the sea. To understand what is happening to the oceans around us, monitoring of oxygen level is also an important requirement. The oceans are being detrimentally impacted by anthropogenic and environmental factors.

Thermal Pollution

Another type of pollution is thermal. The definition of thermal pollution is 'the harmful release of heated liquid into a body of water, or heat released into the air as a waste product of business'. This refers to the thermal pollution caused by industries and businesses drawing water from nearby lakes, oceans or rivers, heating it, or cooling it, before releasing it back into the environment.

This process is detrimental to local communities and ecosystems, and the problem has been increasing over time. One common cause of thermal pollution is when a facility or plant draws water and drastically cools it to clean machinery or other products, and then releases the water back. Wastewater facilities and production plants are the most significant cause of thermal pollution, and the effects of careless use of water have had a tremendous and lasting impact. Even though there are government norms for temperature of discharged water, these are regularly flouted and power plants in particular tend to discharge large quantities of warmer water.

Warmer temperatures decrease Dissolved Oxygen (DO) levels in water and rising temperatures suffocate marine wildlife and plants. Other effects are growth of algal blooms further reducing oxygen levels. The ecological impacts of warmer water being released into natural water leads to thermal shock and is detrimental to many ecosystems. Congenital disabilities and premature release of eggs can result, leading to severe abnormalities in marine life. When temperatures increase, some organisms try to migrate to environments that are more hospitable. This mass migration can lead to disturbances in the food chain.

Power plants use large amounts of water drawn from the sea. Waste heat is dissipated into the environment by using cooling towers, which are cooled by surface seawater at ambient temperature. However, often this is still inadequate and temperatures are higher than desirable. An important issue is how this warm condenser seawater can be discharged back without harming the environment.

An efficient way to utilise the heat available in the condenser reject water is to produce fresh water simultaneously to cool down the seawater before discharging it. The plant puts waste heat into good use and the generation of fresh water reduces the load on local water sources.

One of the effects of rise in temperature, either due to climate change or due to thermal pollution, is coral bleaching. Corals under heat stressors can become white due to loss of symbiotic algae and photosynthesis pigments. Coral bleaching and subsequent loss of coral coverage can result in decline of fish diversity. Coral reefs act as a habitat for sea life and support some of the most bio-diverse ecosystems on our planet. They also act as natural barriers that absorb force of waves and storm surges, keeping coastal communities safe. They support organisms at the base of ocean food chains. Coral reefs cannot afford to be lost as they are key to maintaining marine life diversity.

Effect of Microplastics

An important ocean pollutant is plastic. Apart from plastic debris from dumping waste, microplastics are causing severe harm to aquatic life. They can result from large plastic debris breaking down, or microbeads. Microbeads from personal care products can pass through filtration systems and end up in the ocean. Effluents from waste water treatment systems also contain microplastics. Apart from releasing chemicals they can be digested as food and this has toxic and mechanical effects, including reduced food intake, suffocation, behavioural changes and genetic alteration. Chemicals in microplastics affect humans by changes to genetics, brain development and respiration rates. We need to address all these anthropogenic impacts and work towards sustainable technologies to mitigate climate change.

Sustainable Technologies

It is evident that the ocean is a huge storehouse of resources, has its own delicate ecosystem and needs nurturing. Today use of renewable energy forms in lieu of fossil fuels are being sought to reduce greenhouse gas emissions, which are a large contributor to global warming and therefore, climate change. While solar and wind energy are popular forms of renewable energy, they have largely been employed on land. Offshore wind is also being considered. The vagaries of the ocean make it challenging. In the Indian context, the long coastline, and being a peninsula with a huge EEZ, the need to harness energy from the ocean is prudent. Waves and currents are possible sources of energy. The world over, particularly in northern latitudes in the European continent, marine energy has progressed considerably, especially in wave and tidal energies. The Ocean Energy Systems Technology Collaboration Program (OES TCP) under the IEA has recently released a vision document for marine energy where the number of European countries adopting it is promising [4]. India is still in infancy in this sector and has currently been successful in demonstration projects at small scale.

Why is it difficult to harness ocean resources like energy? Devices to harness wave energy have to meet challenging goals like converting wave motions to energy continuously, when the waves are themselves of different magnitudes and time periods. They are also seasonally directional. The device for it has to perform in ever changing wave conditions, and also structurally withstand severe environmental conditions like cyclones. Various technological options have been, and are being tried, to harness wave energy. Many have worked in field demonstrations but they need to be scaled up and made commercially viable.

The northern latitudes have very high wave power density and thus sustainability of devices in bad weather is a serious challenge. In tropical countries like India, closer to the equator, the wind and wave intensities are much lower. Thus, safe designs for extreme weather are possible, but commercial viability is possible only at large scales. The National Institute of Ocean Technology (NIOT) is the only institute in the country developing technologies for the ocean. The type of devices being studied in India are Oscillating Water Column (OWC) and point absorbers. NIOT has developed devices based on OWC principle and a small device, which is the wave energy powered navigational buoy has been developed and transferred to industry. Similarly, a hydrokinetic turbine using current velocities in tidal streams has been developed as an off-grid device. These have special utility in remote locations like Andaman and Nicobar Islands. Harnessing larger power is capital intensive and owing to perceived risks offshore, there is a reluctance to invest.

While waves and tides are varying, a baseload form of power is possible from the temperature gradient in the ocean between surface and deep sea. In tropical countries like India, the surface temperature is warm throughout the year and can be used to generate continuous power. NIOT has developed an unique technology to generate fresh water from sea water using the ocean thermal gradient. This technology has been very successful in several islands of Lakshadweep. These plants use diesel generators for their power requirements. In an effort towards clean water and green energy, NIOT is now attempting the first ever OTEC powered desalination plant which, when operational, can be off grid [5]. This will be a first step towards diesel displacement in such plants. Scaling up of these technologies require going offshore in deep water. Designs are being developed and studies conducted to determine viability. Alternate markets are being explored for ocean energy.

There is no doubt that ocean energy has a huge role to play in the future scenario of renewables. While the ocean can play a role in clean energy, other impacts require mitigation through sustainable methods.

The oceans are a complex entity filled with uncertainty, both positive and negative. Human intervention has not always been successful. For instance, a breakwater to create tranquility for a harbour is a challenge to design and install, since it has to withstand a 100-year storm. Many anthropogenic activities are seriously affecting the oceans today. Green energy is pivotal to mitigate climate change which will further reduce temperature effects, ocean warming and sea level rise.

Humans are the main reason for all forms of pollution affecting the ocean today and this needs to be understood and addressed. The effect on biodiversity is already being seen, and if not contained, the delicate balance of marine life will be destroyed, eventually affecting land-based flora and fauna. Mankind will continue to explore deeper depths but any interventions need to be sustainable and keep the ocean safe in its pristine state and grandeur. Sustainable technologies to harness resources in an environmentally friendly manner are the need of the hour. We must commit to protect our oceans while using them in a sustainable manner utilising green technologies.

References

1. TIFAC report, Water, Global Technology Watch Group. Department of Science and Technology, 2019.
2. Deepak Samuel V, Purvaraja R, et al "Vulnerability of coastal ecosystems and habitats due to erosion", Climate change and the vulnerable Indian coast, MoEF, November 2018.
3. Mohan R, Kankara R.S. and Venkatachalapathy, "Oil spill projecting modeling of Chennai coast, east coast of India, ", INCHOE 2014 proceedings, February 2014.
4. https://www.ocean-energy-systems.org/ocean-energy/international-vision-for-ocean-energy/
5. P.Jalihal, Energy and freshwater in the context of blue economy, Current Science Special Section, Volume 126, Issue 02 p. 0185, 25 January, 2024.

CHAPTER 18

The Philosophy of Water

Mridula Ramesh

"நீர்இன்று அமையாது உலகெனின் யார்யார்க்கும்
வான்இன்று அமையாது ஒழுக்கு.

"As the earth lives off the sky's bounty,
rain is praised as the nectar of immortality."
~Thirukkural, Arathu Paal, Kural 20[1]

1. Pope, G.U., Drew, W.H., et al. Thirukkural (with English Translation and Commentary). London: W.H. Allen & Co., 1886.

Mridula Ramesh

Mridula Ramesh is the award-winning author of the critically acclaimed *The Climate Solution* and *Watershed,* besides being the founder of Sundaram Climate Institute, which focuses on waste and water solutions. She is an active cleantech angel investor with a portfolio of over thirty start-ups, and is involved with several initiatives to spur climatetech entrepreneurship.

A graduate, with distinction, from Cornell University, with an MBA from the Kellogg School of Management, Mridula worked at McKinsey in Silicon Valley and is the executive director of Sundaram Textiles – the company that was amongst the first textile companies globally to win the coveted TPM Excellence Award. She is part of the board of trustees of World Wildlife Fund, India; and former chairperson of the board of governors at National Institute of Technology, Andhra Pradesh.

She lives and experiments in Madurai in a nearly net-zero-waste home, with her husband and two children.

The Philosophy of Water

Mridula Ramesh

What is Philosophy? It's a portmanteau term from the Greek words, 'Phile' – to love, and 'Sophia' or wisdom, signifying a love of wisdom. a theory or attitude that acts as a guiding principle for behaviour[1]. So, when discussing the philosophy of water, we are looking at the wisdom of water, or more pragmatically, answering questions like:

- Who owns water?
- Who has a right of water?
- Are there limits to the rights?
- Who manages water?
- Does water have value? Should that be reflected in a price?
- Should everyone pay a price? The same price?
- Do all humans have a right to water? An equal right? At all times?
- Do non-humans have a right? If not, why not?
- Does water itself have a right?

As you can imagine, answers to these questions shape how water is perceived by society with very real consequences, for both the water of a region and the fate of that society.

1. Definition of 'philosophy' as stated in Online Etymology Dictionary.

The next question becomes, what is the water philosophy of India today?

Our Philosophy Today

If we walk through the questions above, we see that, paradoxically, water, while ubiquitous, has become invisible, owned by all, with paper-rights given to all, but in practice, available more for the powerful than the disadvantaged. Communities across India are starved of water in the cruel height of summer, while deluged in the rains. Water has become valueless, thus causing a clearance of forests and lakes across India, and the cultivation of water-hungry rice in Punjab and sugarcane in Marathwada. The poor, who act as the fig leaf behind populist treatment of water, end up paying an exorbitant price for water, in their time, their health and their dignity. Because water does not have a meaningful value, it is not managed, or at the very least, not managed well.

I will not dignify the question on non-human right to water with a response.

This philosophy ensures we bounce from drought to flood and back again, as the warming climate speaks most forcefully through water. Especially with a water system like India's which is highly seasonal, geographically varied, temporally skewed, where the rainfall varies greatly annually.

The next set of questions begin with, have we always been this way? No, we have not. Indeed, if we parse the clues of more traditional water engineering, the treatises on water management, and even archaeological and literary evidence, a picture of quite advanced and sensible water management emerges. So, what changed? And why?

To answer those, let us go back five thousand years to the Indus Valley Civilisation.

Our Earlier Philosophy

As we are slowly discovering, the Indus Valley Civilisation may have well been one of the largest ancient civilisations that ever existed, occupying an area larger than the Egyptian and Mesopotamian civilisations combined.[1] It covered large parts of modern-day Gujarat, Rajasthan, Punjab, Haryana, parts of western Uttar Pradesh in India, and Punjab and Sindh in Pakistan. The availability of water varied widely: some sites were very dry, getting less than 500 mm of rainfall a year, while others were much wetter, and additionally were located close to perennial water sources. The Indus residents, or Meluhhans as we now believe they called themselves, adapted to this variability by changing the design of their cities and matching crop to water.[2]

Dry places saw farmers primarily grow millets, while towns and larger cities in dry places, like Dholavira, are defined by reservoirs to store water. Water-abundant places like Harappa, that enjoyed the waters from the Ravi and Beas rivers, saw their farmers growing two crops: a winter crop of wheat or barley, and kharif crops of millet, cotton or even rice. Mohenjo-Daro, which lay close to the mighty Indus, could afford to be profligate with its water, even sported giant baths.[3]

Contrast this with our modern cities – Bengaluru, Delhi, Mumbai, Chennai – that do not vary their design to suit water availability; they try to reshape the water, fetching water from several hundred kilometres to quench their thirst, to suit their design. The consequences of that philosophy are visible in the

1. Petrie, C.A., Singh, R.N., et al. 'Adaptation to Variable Environments, Resilience to Climate Change: Investigating Land, Water and Settlement in Indus Northwest India'. *Current Anthropology* 58, no. 1, 2017; Weber, S. 'Seeds of Urbanism: Palaeoethnobotany and the Indus Civilization'. *Antiquity* 73, no. 282, 1999; Personal communication with Cameron Petrie in 2021.
2. Ibid
3. Ibid

flood-and-drought bipolar reality of these cities today. Farmers today pump water from deep underground to grow rice and sugarcane in dry places, forsaking crops that would thrive in them. This mirage-like-security will hold as long as the groundwater does. And as that shakes, the house-of-cards food security will come crashing down alongside.

Water was respected and managed, and the need for everyone to do their bit to maintain waterworks was further reinforced by good storytelling. I come from Madurai, which revolves, literally and figuratively, around the Meenakshi Amman Temple, the home of the 64 stories of divine sports.

In one story, the Vaigai, Madurai's primary river, was in spate. The king ordered each family to help build a protective bund around the city. One old woman, a *puttu* (or rice cake) seller called Vanthiammai, offered puttu to anyone who would help her, as she lived alone, but no one came forward. Lord Shiva in the form of a young man offered to help her but after eating the puttu, fell asleep. The king's guard discovered him sleeping and he was whacked with a cane. The story goes that every citizen of Madurai felt the blow. A festival celebrates this story every year and reminds us that no one is exempt from their responsibility in water management.

Ancient political strategist, Chanakya, adds his bit. He understood that a king's power rested on prosperity which depended on good harvests. Crop yield depended on water, and so he decreed: 'All water belongs to the king and users will pay a water tax for taking water from irrigation works built by the king.'[1] Paradoxically, we've not heeded his diktat, while modern-day Israel and Singapore, poster-children of good water management, have.

1. Kautilya. *Arthashastra*. New Delhi: Penguin India, 2000.

Chanakya further understood that variable water as we have in India, had to be managed well, and therefore had to be fairly valued. So all water, including water used for agriculture was priced, and while everyone paid, the rich paid more. So those who used technology to transport more water, typically the wealthier farmers, paid more tax as a higher share of their crop.[1] Today, the situation is reversed. Wealthier households and farmers, with access to borewells that run on free electricity, can draw water easily while poorer farmers sometimes pay them to access water. In contrast, Israel and Singapore provide a basic water allowance at a low cost, with progressively higher tariffs applied to consumption beyond that level.

Chanakya also gave tax-breaks for maintenance – sometimes equivalent to building new irrigation structures. Importantly, taxes were calculated as a share of the crop production: during years with bountiful rains and harvests, the tax was higher, while in drought years, tax was automatically lower.[2]

That changed with the British.[3] They imposed a cash tax on the farmers. Rain or no rain, a farmer had to cough up the same amount of tax in cash. And so, farmers turned to cash crops, those that had a ready market and therefore could easily be sold, like indigo, sugarcane, wheat and cotton.

Forests were cleared across the length and breadth of the country to create 'new' land to grow cash crops: sugarcane replaced *dhak* forests in Uttar Pradesh, wheat and cotton replaced millets in Maharashtra.

Then there were the dams. Before the British, and certainly in ancient India, the primary method of irrigation were the tank systems, which were often a system of cascading lakes created

1. Ibid
2. Ibid
3. The section on the British is based on Ramesh, M. *Watershed*: Hachette India, 2021.

by understanding the drainage patterns of an area. These were community managed and decentralised in function.

The British, however, prioritised a philosophy of centralised control, where the ruler could quite literally turn off the switch on dams, creating geopolitical weapons of power that have relevance even today. If you doubt this, look at the dams on subcontinental cross-border rivers and the chatter enveloping them.

Many times, the new crops did not sync well with existing rainfall patterns. And the clearance of forests altered the climate and river flow. And so, when the rains failed, as they do every few years in India, the farmer fell into debt, and famine visited the land.

This fascination with 'modern' technology, and the belief that it could solve any problem continued post-Independence. In the aftermath of the 1965-'66 famine, when India had to plead with US for wheat, the leaders placed their faith in the borewell, which powered the Green Revolution.

The Green Revolution made modern India food secure. But there is always a price. The northwest of India is dry, getting about 500-600 mm of annual rainfall in most places.[1] Rainfed rice needs between 1200-1400 mm in its growing season[2], and so the northwest makes up the difference with groundwater. Recently, scientists found that globally, groundwater extraction for irrigation is changing Earth's rotational tilt[3]. The fact that we have not noticed it is a telling sign of our water philosophy: water is invisible, especially when it comes from underground. But

1. Water Resources Information System, Ministry of Jal Shakti, Government of India.
2. Maiti, A., Hasan, M.K., Sannigrahi, S. et al. Optimal rainfall threshold for monsoon rice production in India varies across space and time. *Commun Earth Environ* 5, 302 (2024).
3. Seo, K.W., Ryu, D., Eom, J., Jeon, T., Kim, J.S., Youm, K., Chen, J. and Wilson, C.R., 2023. Drift of Earth's pole confirms groundwater depletion as a significant contributor to global sea level rise 1993–2010. *Geophysical Research Letters*, 50(12), p.e2023GL103509.

when you suck up so much water from under the ground, levels fall, and so a law to conserve groundwater pushed the sowing date for paddy in Punjab. And when that happened, the gap between paddy harvest and wheat sowing shrank. And farmers turned to the cheapest and fastest way to clear their fields – they set the stubble on fire.

The borewell and groundwater reshaped cities as well. Anyone who could, drilled underground and cities expanded, no longer constrained by the municipality's ability to provide water. Municipalities did not cover their costs with the water tariff and failed to collect much of that paltry amount. Consider this: In 2018-'19, the Brihanmumbai Municipal Corporation (BMC), Mumbai's rich municipal corporation, spent ₹19.44 per kilolitre in to source, transport, treat and distribute water to Mumbai residents.[1] But the corporation charges only ₹5.22 per kilolitre for non-slum households who use up to 750 litres per day. This means a family of five pays less than a few plates of pav bhaji for their monthly water use. And it's not just about protecting the vulnerable. In October 2020, a proposal to raise the tariff for those using more than the minimum level of water was blocked by corporators across party lines.[2]

Why?

Competitive politics had entered the realm of water. Water had become a largesse for politicians to provide, and with that, the need to manage water was given the go-by. And when that happened, lakes across Indian cities began to disappear.

Sundaram Climate Institute, which works in climate resilience, crowdsourced groundwater level data from thousands

1. 'Corporators across Party Lines Block BMC Proposal to Hike Water Charges'. *Indian Express*, 2020.
2. Ibid

of people around Madurai's tanks between 2018 and 2021. When we collated those responses, we found that tanks appear to support groundwater levels in their vicinity, keeping levels 200 feet above what they might have been without a tank (see Figure below)[1].

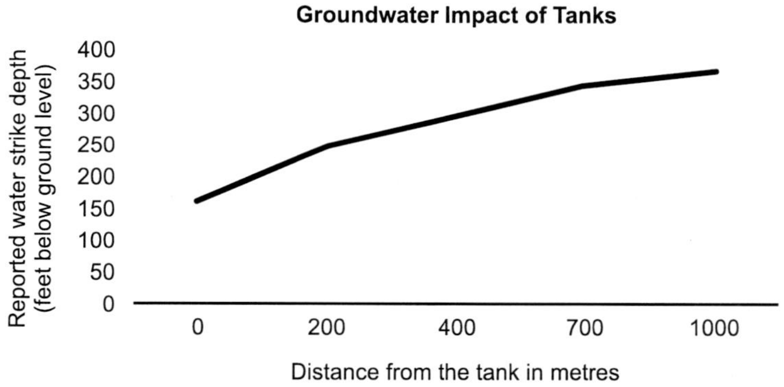

This increase in ground level water literally acts as gold in a drought. But the service is invisible, and when the cost of drawing up groundwater is miniscule or free, no one values the service as they ought, until, of course, the water runs out. And that's precisely what is happening in the peripheries of so many Indian cities, in the pre-summer and summer months, especially in dry El Nino years. Parse the wails of people complaining of no water during a crisis, the voices of communities living around a lake will be silent.

And when it pours, which is a feature not a bug of the Indian monsoon, with no lake to catch the downpour, it floods. In Chennai, many of the neighbourhoods that were most impacted were either lakes or held a giant lake, which had morphed into offices and homes.

Even as the structural resilience of our cities fall, the financial muscle of the corporations to deal with climate impact

1. 'Making India Water Resilient – Lessons from Madurai'. Sundaram Climate Institute, 2023.

is weakening. Apartments in Bengaluru pay far less for piped water than households in cities in Sub-Saharan Africa and Latin America.[1] Where will the investment to treat sewage come from? How will the salaries of staff be paid who will audit builders to check if there is violation of norms? So violations go unpunished and so it remains under-treated, depriving us of a valuable source of water in a drought.

A warming climate accentuates the contrasts of water. It makes wet regions wetter, and dry regions drier. It makes the wet seasons wetter and dry seasons drier. It reduces the number of days on which it rains, and increases the frequency of extreme events. And by doing that, it is shining a light on our water philosophy.

And the emerging picture is not pretty.

So what next?

Paraphrasing a dialogue form Christopher Nolans, 'A Dark Knight', what happens when an unstoppable force meets an immovable philosophy? You get floods and droughts, often in the same year in the same city. Year after year after year.

Something has got to give. And I don't think it is going to be the warming climate.

1. Mitlin, D., V.A. Beard, D. Satterthwaite, and J. Du. 2019. "Unaffordable and Undrinkable: Rethinking Urban Water Access in the Global South." Working Paper. Washington, DC: World Resources Institute.

“A drop of water, if it could write out its own history,
would explain the universe to us.”

~Lucy Larcom

CHAPTER 19

The Water of Consciousness

Minni Jain and Philip Franses

"I knew then that "w-a-t-e-r" meant the wonderful cool something that was flowing over my hand. That living word awakened my soul, gave it light, hope, joy, set it free!"

~Helen Keller

Minni Jain

For more than 30 years, **Minni Jain** has been working with communities to regenerate their lives and landscapes. As co-founder and operations director of the Flow Partnership, she works to spread community-led, simple, successful, low-cost, traditional wisdom and methods of holding water and managing floods and droughts. She has helped set up practical Water Schools in Africa, India and Europe that operate both as online learning forums as well as on ground community water hubs. Minni was born and brought up in the Himalayas in India and now lives in the UK. Minni is co-author of *The Language of Water* (Synergetic Press, 2025)

Philip Franses

Philip Franses uses his expertise in holistic thinking and teaching to address global challenges through multi-stakeholder processes. He studied mathematics at New College, Oxford, UK and has designed intelligent software for a variety of organisations. Philip is the co-founder of The Flow Partnership, through which he creates platforms for communities to share their knowledge with each other, helping them restore water to its vital place within the cycles of nature. Philip also teaches Holistic Science and is co-author of *The Language of Water* (Synergetic Press, 2025)

The Water of Consciousness

Minni Jain and Philip Franses

"Climate breakdown or global warming is NOT an ecological problem."

With that in mind, while you ponder on it and gather arguments for or against, let us first dive into a project '*ReForesting Baran*', that we are working on in a very dry and arid part of the Indian states of Rajasthan and Madhya Pradesh (MP). The project is regenerating the ecologically degraded and almost dead periphery of the Kuno National Park, a tropical forest in Rajasthan, bordering MP.

ReForesting Baran

The Kuno forest has shrunk a bit over the last 20 years and for seemingly endless miles around it, the land on its periphery looks like dry, ecological wasteland. Hot and dusty, with hardly a few scrubs and bushes and the odd tree growing here and there, very little looks alive there. One wonders how anyone living there could have any food, livelihood or life.

One of the primary reasons for this degradation seems to be the displacement and resettlement of the Sahariya tribals, from inside the Kuno forest to its outer edges. In 2010, 23 villages (1547 families) within the forest were resettled outside to make way for the endangered Asiatic Lion, other than at the Gir Sanctuary in the state of Gujarat.[1] For various reasons no lions have been moved from the Gir Sanctuary to Kuno Forest and instead, in 2022, 8 cheetahs from Namibia, Africa were flown in and resettled there with mixed success.[2] This project aims to bring back cheetahs to India, where they have now been extinct for over 70 years – an ever-evolving big cat resettlement chequerboard meant to benefit humans and cats.[3]

In the meantime, the Sahariya tribe, who were also known as guardians of that forest, were resettled from Kuno onto the peripheral lands outside. They have had to re-wire their livelihoods: to become foragers and farmers in a land that is increasingly devoid of any ecological health – no water, no trees, no fertile soil. The state of these Sahariya tribal people today is worse than pathetic. There is rampant distress migration, trafficking, illness, disease, malnutrition and hunger amongst them to name just a few of the issues they are contending with.[4]

We, The Flow Partnership and Parmarth India, along with the communities, mapped 520 square kilometres to revive the water in all the catchments there under a project with the name '*ReForesting Baran*'. Beginning in 2024, we constructed the first 5 water bodies, holding a combined 122,390 cubic meters of water that was otherwise flowing away and lost from the local water cycle. Pre-monsoon the land was completely dry and brown. When a water body was built in the area to hold the monsoon rainwater, it brought life, vegetation, agriculture, and diverse harvests including tomatoes and fish! A thriving landscape has started to return. From the time taken to construct the water

1. Dutta – see refs
2. Biswas
3. National Tiger Conservation Authority
4. Ghai

body, to the monsoon rain and post monsoon, water availability for the villagers for the entire next year is changing the whole dynamic of the region. Done at catchment level with multiple water holding strategies (primarily staggered pond building and planting of hundreds of trees and looking after the health of the soil) the revival story is truly eye catching. We see landscape magic happening as quickly as in 4 months – from dry and brown to lush and green.

The Sahariya tribe are now beginning to have the wherewithal to tackle their problems at their root, as well as to remake a connection to a healthy life outside the forest.

What is at the heart of this feeling of reconnection?

Unity

When Gautam Buddha, "freshly enlightened", was traveling to Sarnath where he delivered his first sermon, a shepherd noticed the halo of radiance around the master, and bedazzled as he was, he asked, stupefied– "What are You"?!

Notice, he exclaimed 'What', and not Who.

"Are you God"?

"No", said the Buddha.

"So are you an angel"?

"Not that either", remarked the Buddha, in his characteristic calm demeanour.

Frustrated, the shepherd remarked– "Then you must be a human being".

"I am not even that. I am Buddha, the awakened one".

In India, we have access to many spiritual traditions. In some traditions, especially the Jains, all living beings are given the same status of having a soul and there is no difference at all between the observer and the observed. Not only are the two interconnected, but the journey of experiencing the single coming together of the observer and the observed, *is* the human journey to consciousness and enlightenment. Such glimpses of this unity are few and far between and we understand that one can spend many lifetimes chasing those glimpses that occur more and more frequently till one joins the 'awakened ones' as the Buddha so eloquently put it.

What is consciousness and where does it arise?

In our materialistic world today, to work with water is seen as the job of the engineer. But traditionally, the community had quite another relationship with water. Their spiritual beliefs and myths carried tales of how to hold water in their lands and lives. Water made the human, and the human was respectful of how they related to water.

The universe manifests itself physically and as conscious experience. Consciousness is generally understood as the 'awareness of self' and this has, of course, been the subject of debate between humans of all persuasions across millennia. Does it arise in the mind or the body? Or does it exist somewhere else? This is a question that has been puzzling the West, and the East, for many centuries. While there are great strides in understanding where it arises, not the least with huge influence from Indian sacred texts, we are aware that consciousness can only be experienced. If you say it is the mind that allows you to experience consciousness then you limit your perception of ultimate reality to a fraction of its nature, in as much as the limits of your mind can help you understand. Yet, consciousness is an infinite vastness that we have glimpses of from those who have gone towards it before us, or through self-exploration routes like meditation.

Within all these debates of where it arises or abides, a central fact often gets obscured or missed: you have to be ALIVE to understand, feel or experience consciousness.

All the major religions of the world give us a sense of the vast sea of consciousness (Unity/ Bliss/ Oneness/ Pure Knowledge/ Paradise/Jannat), that a tradition may choose to describe or find paths towards.

The Jains, the Hindus, the Buddhists, the Christians, the Muslims, the Sikhs, the Jews, all have their own language often based upon a trinity, postulated in different words. From all of them we get a route to access consciousness. It's a push and pull between matter and mind and a transcendence of both to reach the state of supreme unity/ knowledge/ paradise/ heaven/ moksha or enlightenment – whatever one may choose to call it.

From the Jains we learn right perception, right knowledge and right action. Unpacking that a bit, what is right perception? That with which we can cut through all illusion to see the true nature of the Universe.

What is right knowledge? Dropping illusion, understanding impermanence, knowing the true nature of reality and the structure of the Universe.

And what is right action? That action arising in a merging of the observer and the observed, the action that holds all of potential.

Right action is that which we have to take *together with* right perception and right knowledge.

This wisdom has formed the basis of most religions and spiritual traditions and one can access any number of writings to go deeper into these and also find the parallels for our own deeper understanding of them.

They are the touchstone on which a view of a balanced planetary ecology rests.

Consciousness is not out there, it is within us. The capacity to have experiences is within each one of us. The agency to think and take action is within us. Endless discussions alone cannot solve the climate problem, endless optimism and hope alone cannot change reality. And it is the lessening and absence of right perception, right knowledge and right action which is leading us to the current crisis of planetary breakdown we are witnessing today.

Fragmented quantity over qualitative unity

The history of communities being guardians of the water is long. Yet modern man tends to relegate it to being something very minor or 'sweet' or 'traditional' and tends to talk of it as if it was nowhere near as important as the studies that are conducted by scientists measuring the degrees of global warming.

From the scientist's point of view, it is useful to conceive the world through the lens of abstract terms, such as referring to water as two molecules of hydrogen and one of oxygen (H_2O). H_2O makes water a subject in a way that everyone shares the same basis of understanding of it. Water ceases to be a quality or a life giving force and instead becomes a conceptual formula. If we want to know anything about water, we have to appeal to the expert scientist, and if we want to change the behaviour of water, we need to ask the engineer.

Abstract pictures that we create in our minds and the laws that regulate behaviour only work if they oversee all aspects of phenomena. In the case of water, this conceptualisation into H_2O falls far short of this objective. In the moment during and just after a raging flood devastating a village, or when a particular hot summer takes away all water from the land, to see water as a collection of H_2O molecules is a completely redundant and unhelpful picture. In such moments of crisis, the villagers and communities are confused and disempowered by the imposition that water is something they need the help of an expert to address.

In regeneration projects, the first thing that needs to happen is to re-empower the communities with *their* story of how they can return to their traditional knowledge of how to hold the water in the landscape. In times past, the community treated the water with respect and had their own methods of how to hold the water in the land to lessen floods and droughts. That was the norm not the exception. The communities had water embedded into the context of the landscape and their lives. They knew not to cut down the trees, to tend to the soil, to build structures to manage the flow of water if life was to be harmonious. This knowledge of holding water in the landscape was passed on from generation to generation. It is only recently their worldview has begun to be overwritten with one of measurement and control.

The importance of quantitative measurements that tell us the level of degradation we have on the planet today are of course critical in informing action. However, they are the end game. Unfortunately, numbers and quantitative measurements and their importance has superseded the cosmic understanding of water's role in life. The start of the game is in knowing where the problem arises - in our consciousness and in the separation we experience in the observer and the observed.

The basic and essential material foundation of life (and by inference for the human level of consciousness to manifest) is water. This is not abstract or inferred or imagined – this is the reality.

In times not so long past, in India and indeed in the rest of the world, the guardians of the water were called *gajdhars* or holders of measurement.[1] The gajdhars did not just measure where to build a pond and its dimensions, they roamed the land and at a given point they would point and say where to build the particular pond in relation to the whole landscape. What went on before that pointing to the right spot for the water body in the landscape, is the glimpse into the water of consciousness and can only be experienced as a whole, and then, only then, can it be articulated in numbers and rule measurements.

Numbers too of course represent a divine unity that is not to be separated from just a finite value. Here we are reminded of the Indian mathematician, Srinivas Ramanujan, who was given the fellowship in Cambridge and is regarded as one of the greatest mathematicians of all time, though he had almost no formal training in pure mathematics. His "lost notebook", containing discoveries from the last year of his life, caused great excitement among mathematicians when it was rediscovered in 1976 and some of these formulae are still being worked out.

He often said, "An equation for me has no meaning unless it expresses a thought of God."

And so with the gajdhars. They had an implicit understanding of unity in measurement, life and landscapes. Gajdhars were given the respect and acknowledgement of being those who could also measure the depth of the time, the era, the society in addition to being able to measure the dimensions of a water body.

Today, globally, while expertise in measurement is unparalleled, that unity needed in measurement has turned into nought. The depth of our cosmos/planet/time/society/era is excluded in the measurements of the water body to be built. The access to the divine in measurement or the vast unity of consciousness and the true knowledge of the universe that

1. Mishra, p.19-20

underpins the flowing form of the water body, is lost in the design.

The expertise of the gajdhar, or water diviners in other cultures, has been lost or relegated to the mere position of a 'new age' quack living on the fringes of society and 'who knows how much of what they say can be believed'.

Why has mathematics and science become separate from reality?

The relation of observer to the observed is not foreign to science. Indeed, in quantum theory the paths of western science and eastern wisdom were close to meeting. Experiments into the elementary nature of matter were only able to conceive of elementary reality in the way the relationship of observer to observed presented itself to measurement.

"In classical physics, science started from the belief – or should one say the illusion? – that we could describe the world or at least parts of the world without any reference to ourselves. [In quantum theory] what we observe is not nature itself, but nature exposed to our method of questioning."[1]

"The quantum postulate implies that any observation of atomic phenomena will involve an interaction with the agency of observation not to be neglected."[2]

Such was the sway of a material picture of reality on science, that quantum theory tried to hold on to the classical picture of material reality, even though the lens through which the particle world was depicted was unmistakably the relation of observer to observed. Quantum scientists like Neils Bohr and Werner Heisenberg saw resonance in their work with Indian scripts about the observer and the observed. Here was the great chance for

1. Heisenberg, p. 43, 46
2. Bohr, p.54

western science to see that all reality stems from how the relationship of observer to observed unfolds into a landscape of perceived existence. But western science could not go so far as to say that the navigation of the observer to observed was the domain of choice of right perception, right knowledge, right action. The quantum scientists reversed this eastern view by saying it was mathematics alone that was able to organise the instance of observation into a sum of predictable behaviour. In quantum physics the limitation of individual knowing is taken as an impediment that prevents it entering the world that mathematics depicts.

When we think we are being scientific and reducing an environmental challenge to numbers, such as the degrees of global warming, we are concealing the true nature of our engagement with a reality that creates itself dynamically in the relation of observer to observed. Today quantity and measurement has overtaken every aspect of life to the exclusion of any other quality or connection life may have.

Quantum entanglement is another example. At life's most basic level, when two particles, such as a pair of photons or electrons, become entangled, they remain connected even when separated by vast distances. When one observer makes a measurement of one particle, the entangled particle is simultaneously revealed as possessing its complementary property.

And so it is with the planet. When we hold water and regenerate a physical landscape then we feel the path of right perception, right knowledge and right action in the return of the area to life. We see, in this act of restoring a local environment, that we could apply this same process to the whole planet. At a microscopic level regeneration of part of the planet affects the planetary water cycle too. We are all connected in our actions through the consciousness of right behaviour.

Here is the AI explanation of the observer and the observed: "the observer (the "I" or ego) and the observed (the world and our experiences) are not separate but interconnected. Essentially, what we perceive as the self, the "observer," is actually a product of the experiences it is observing, and therefore the observer is also the observed."

Water is the medium by which our individual actions and those of the community scale up through our consciousness to see the path of right behaviour at the scale of the whole planet.

Climate breakdown is actually the breakdown of connection between the observer and the observed in a loss of right perception, knowledge and action.

Conclusion

Only in being alive can we explore consciousness. If there is no water, then we are not alive and with it disappears the chance for us humans to understand our purpose in life of right perception, knowledge and action in this cycle of cosmic time. Only if we have life can we take right action in an 'awakened' humility of right purpose. And for life to exist we need water.

We end here with a simple message drawn and written on the walls of a water body in Bhilwara, Rajasthan by an unknown tribal artist to remind us of the unity of life and its ultimate interconnectedness with water:

"The centre of life is water. These are the beautiful waves and these are the stairs to get to the water and these are the trees and these are the flowers which add fragrance to our lives. This is the message of the desert and its people."[1]

1. Mishra, Aaj_Bhi_Khare_Hain_Talaab

References

1. Biswas, S. (2023) Cheetahs in Kuno: Is India's Effort to Reintroduce the Big Cat Facing a Crisis? – accessed online at https://www.bbc.co.uk/news/world-asia-india-66392483
2. Bohr, N. (1987a) The Philosophical Writings of Niels Bohr; Volume 1; Atomic Theory and the Description of Nature; Ox Bow Press
3. Dutta, A. (2019) Kuno Indias Second Home for the Asiatic Lion is Reary (2019) Mongabay accessed online at https://india.mongabay.com/2019/11/kuno-indias-second-home-for-the-asiatic-lion-is-ready/
4. Ghai, R. (2025) Project Cheetah in Kuno National Park Not Environmentally Just – accessed online at https://www.downtoearth.org.in/wildlife-biodiversity/project-cheetah-in-kuno-national-park-not-environmentally-just-paper
5. Heisenberg, W. (1958) Physics and Philosophy; Penguin Books
6. Jain, M. and Franses, P. (2025) The Language of Water; Synergetic Press
7. Mishra, A. The Ponds Are Still Alive And Relevant Today (original title Aaj Bhi Khare Hain Talaab); Gandhi Peace Foundation – accessed online on 5th October 2023 at https://www.indiawaterportal.org/articles/aaj-bhi-khare-hain-talaab
8. Aaj_Bhi_Khare_Hain_Talaab – accessed online at https://commons.wikimedia.org/wiki/File:Aaj_Bhi_Khare_Hain_Talaab_%28Hindi%29.pdf
9. National Tiger Conservation Authority – https://ntca.gov.in/project-cheetah-kuno-np/#project-cheetah
10. Parmarth India website – https://parmarthindia.com/
11. The Flow Partnership website – www.theflowpartnership.org

CHAPTER 20

Sacred Waters – Timeless Wisdom

Romola Butalia

एवेमा मूर्ता अप उपास्स्वेति ॥

"All these forms are water; meditate on it."
~Chandogya Upanishad 7.10.1.

Romola Butalia

Romola Butalia is a writer: an author, a committed conservationist and a journalist.

As an author, her books and essays are primarily about sacred spaces; spirituality and consciousness; different spiritual traditions and the many paths to the Truth.

Her search for wisdom and joy led to a continuing tryst with the Himalaya where she lives a mostly reclusive life. She is a practitioner and teacher in the traditions of the Himalayan yogis, with the given name, Sriji.

Professionally, she has been a journalist with wide experience in print, visual and digital media. She is the founder of Save Kumaon, and supports initiatives to preserve, protect, foster and rejuvenate the natural ecology, traditions, culture and spiritual heritage of the Himalaya.

Sacred Waters - Timeless Wisdom

Romola Butalia

Through time and space, across cultures, religions and regions, water has been respected, revered and worshipped as intrinsic to life, as the very source of life, and as divine grace. It is the means of cleansing, purification, renewal and rejuvenation. In its myriad forms, water is part of a deeper genetic memory.

Varun, Lord of the Waters and of the Skies, has pre-eminence and roots in the pre-Vedic era; regarded as Lord of the Vedic pantheon, representing the concept of *Rta* or self-regulated law and order in nature, emphasising harmony as sacrosanct. Rta is *dharma* that protects all life, based on the principle of oneness of life and the recognition of its interconnectedness. In Puranic mythology, Indra is the Lord of the heavens. His *astra* or weapon is the *vajra* or thunderbolt that commands the life-giving rain that nourishes the earth, helps crops grow, stores precious freshwater in glaciers and wetlands. Through the generations, we never forgot this.

To be worthy of performing ritual Vedic worship, observance of daily *sandhya*, which blends both worship and meditation, is obligatory. This includes purification with water and the symbolic offering of water to the Sun, giver of *prana,* the life-giving essence of the breath. Remembrance of ancestors is accompanied by the offering of water. Without water, neither Vedic nor Tantric worship can begin. Even the *sankalpa* or resolve to worship is undertaken with water, reminding us that water is deeply sacred, it is the witness, and it connects us to the divine. The rite begins with purification of ourselves and the surroundings with water, invoking all the waters of every river and sacred *tirtha* or pilgrim place, inviting all the gods to be present in the consecrated waters. Worship itself begins with the offering of water to the deities invoked.

Amrit or the water of immortality is the holy water with which Sikhs are initiated into the order of Khalsa Panth or brotherhood. This water is synonymous with the One, and in *Gurbani*, the sacred hymns, amrit is repeatedly equated with *Naam* the Name or *Shabad* the Word. *Sarovars* are water bodies near gurudwaras, with their own history and associations, and are symbols of the core values of Sikhism: purity, humility, equality and communal harmony.

Across different schools of Buddhism, water offerings vary in form, but share a common spiritual significance. In Theravāda Buddhism, pure clear water is offered with humility, symbolising the purity and clarity of mind that the aspirant seeks. In Mahāyāna Buddhism, water offerings are part of more elaborate ceremonies and rituals, where they express faith and cultivate detachment to overcome karmas and aspire for one's Buddha-nature. In Vajrayāna Buddhism, bowls of water are offered to purify karma, and aspire for positive qualities like generosity, compassion, mindfulness, detachment and wisdom. In Buddhist cosmology, water along with earth, fire, and air constitute the physical world. Offering water reminds us of the fundamental interdependence of these elements and reinforces our connection to the natural world and all life on earth.

Legend holds that the 24th Tirthankar, Mahavira, attained Nirvana or final emancipation at Pavapuri. Pilgrims visiting the site of his cremation would take handfuls of soil, gradually creating a large pond. Within it, a temple was built. In Jain cosmology, the *Dik Devis* are the guardians of the directions, some of which are symbolically linked to oceans or great rivers. Such associations reflect the enduring significance of water in spiritual and religious contexts, emphasising its role in rituals and the sacrosanct.

Water plays a pivotal role in all Abrahamic faiths, symbolising spiritual awakening and the presence of the Creator. These traditions emphasise the inherent duty to protect and care for water, connecting this responsibility to broader principles of social justice and stewardship.

The parting of the Red Sea by divine intervention is a powerful narrative found in the Hebrew Bible and referenced in the Quran where Moses raises his staff so the fleeing Israelites can escape. When the people were thirsty, Moses was inspired to draw water from the rock at Horeb. "Strike the rock, and water will come out of it for the people to drink" is recorded in the Book of Exodus.

In Judaism, water is a powerful symbol of life, sustenance and purification, used extensively in ceremonial and ritual practices. The book of Leviticus, outlines purification rites and spiritual renewal through water. It is a symbolic element for spiritual rituals, reminding us of the rhythms of nature that sustain and nurture both our physical and spiritual life. It is embedded in Jewish culture and tradition. The Jewish creation narrative begins with water. Genesis 1:2 states, "the earth was formless and empty, darkness was over the surface of the deep, and the Spirit of God was hovering over the waters." Before the emergence of light, water was the foundational element in Jewish cosmology, and is symbolic of creation and fertility.

Water is deeply associated with transformation. Noah's Ark is a powerful reminder that when evil overwhelms the world, it must be cleansed for a new beginning. Moses leading his people across the Red Sea is also the profound journey of transformation from slavery to freedom.

At the same time, the Flood shows that water is both a source of life and a symbol of death and destruction. Perhaps a warning and a reminder that if we move from the natural order that has been ordained, destruction and dissolution is inevitable. If the imagery of creation narratives is of water, creation is followed by cycles of preservation and destruction. It is time that we become more aware of the destruction that we have brought on ourselves by disregarding the time-honoured laws of preservation that connected us to the earth, the waters, and all elements, as well as all living beings.

In Christianity, the ritual of baptism in water symbolises purification and renewal, a rebirth to a life in Christ, guided by the indwelling Holy Spirit. On entering many churches, believers sprinkle themselves with holy water and make the sign of the cross, creating a subconscious connection with holy water and the giver of life, with a deep awareness of the sacred nature of water. We are reminded of this connection through many Biblical stories: how Jesus transforms water into wine as his first miracle, and how he walks on water.

In the Gospels, water is profoundly associated with life and sacredness. Jesus speaks of 'living water' when he requests a Samaritan woman to draw water from the well for him, "If you knew the gift of God and who it is that asks you for a drink, you would have asked him and he would have given you living water" (John 4:10). Water is the metaphor and conduit to eternal life when he says, "Everyone who drinks this water will be thirsty again, but whoever drinks the water I give them will never thirst. Indeed, the water I give them will become in them a spring of water welling up to eternal life." (John 4:13-14). 'Living water' and the 'water of life' are powerful images in Christianity, connected with the life-giving presence of divinity. Living water represents the Holy Spirit as a present, experiential force while the water of life signifies eternity that transcends all mortal needs. Both images are intimately connected with Jesus as the giver of 'living water' now and the eternal source of the 'water of life' to come.

In Islam, which originated in the deserts of Arabia, water is regarded as life-giving, sustaining and purifying. It is considered the origin of all life, from which Allah created mankind. (Al-Furqan Surah 25:54) Cleansing with pure water is essential before the recitation of prayers, a reminder of the role of water in spiritual purification. Water is recognised as divine blessing, and as a source of life, apart from satisfying our physical needs, it is a means for purification and is frequently mentioned in the Quran as a sign of God's mercy. The Arabic word for water, *mā* is found 63 times in the Quran. Islamic teachings emphasise the importance of water conservation and responsible management, reflecting a broader stewardship over natural resources. In a culture deeply aware of the precious nature of water, it is also associated with sharing, universality and charity.

The act of immersion in water is a profound expression of faith and purification, common to most cultures and traditions. Emerging from the water symbolises renewal and is purifying, both physically and spiritually. Millions of people take dips in sacred waters at particular conjunctions of planets, particularly during the auspicious Kumbha, itself the vessel in which *amrit* or the nectar of immortality was carried by the gods as the demons sought it and chased them. The powerful and recurrent imagery that is central to human narratives and conscious living, are a reminder of the deep significance water holds for us.

Different indigenous cultures worship water bodies as dwelling places of guardian spirits of the land and waters, connected with the world of spirits, of ancestors and of deities. Water and its sources were naturally revered as life-affirming. Across different cultures, mythologies closely connect water with existence and creation, reflecting the ancient awareness that water is intrinsic to all life.

With the loss of mythology, and a deep spiritual bankruptcy that overwhelms humanity, the crisis of faith is reflected in our deeper disassociation with nature, with other beings and with both the individual self and the Universal Self. This is part of the barren desertification of the Earth. With the loss of connection with all that was once held sacred, in our self-serving and deeply individualistic goal-setting, we have lost our way home.

Water is closely associated with the purity of the perpetual snow-covered Himalaya and the glaciers from where many rivers take birth. The serene Lake Manasarovar is worshipped as Parvati, daughter of Himalaya and Shakti or manifest energy of Shiva, the Unmanifest, who is revered as Meru or Kailash, the sacred mountain circumambulated by Hindus, Buddhists, Jains and followers of the Bonpo faith. It is regarded as the ultimate tirtha or pilgrimage in these traditions, and undertaken on foot. In the Tibetan tradition, the *kora* is sometimes undertaken with full-body prostration by the devout. Among rivers personified with rich living myths, the Mother Goddess Ganga is deeply revered. The ocean is associated with the *samudra manthan* or churning that released the eternal gems as a blessing, including Sri Lakshmi, benevolent and munificent, the energy of preservation. This churning also caused poison to separate, maintaining the proverbial duality of the manifest world. In many creation myths, creation itself began from the oceans.

The deep spiritual significance of water transcends its physical form. In several traditions, water with its fluidity symbolises consciousness, the ever-changing nature of existence, and the interconnectedness of all beings. Water has long been seen as the mirror that reflected the inner self. Water bodies along with forests, caves and remote mountains have always been considered sacred places for introspection, reflection and deep meditation to understand the nature of existence, the cycle of birth and death, and the mysteries of the Universe. The tides bring home to us the cyclical rhythms of the water cycle and mirror the patterns of life, of birth, death and renewal. We learn to recognise the impermanence of life, and are compelled to be aware of transition and transience as the nature of material existence. By acknowledging our connection with water, by appreciating the respect accorded to water by all cultures, we recognise our inherent responsibility to protect it in its pristine form.

We protect that which we love more dearly than ourselves. We love what we truly know and respect. When did we stop loving the Earth, the Waters, the Fires, the Air and Space, all of which, according to Vedic and other traditions, form our bodies and all matter? When did we stop feeling the pain of all beings? When did we replace the fulfilment that comes from deeper connections with the emptiness of simulation? When will we cease to think of water, and everything including humans, as mere resources? When will we value the intrinsic worth of life and rediscover the harmony that we have lost? When will we measure life through balance rather than in terms of currency?

While ideological differences exist between religions, water is the unique unifying deeply powerful symbol, that we need to value more. The sacred nature of water common to cultures and religions has the power to unite people by reminding us of our shared need for, and responsibility to protect it. Beyond the fundamental necessity for clean fresh water for survival, the spiritual and religious reverence for water can inspire many to deepen their connection with it and support conservation efforts. It does not matter whether one has a secular, religious or spiritual perspective, our collective love and respect for water as the key to life, and this sacred connection can foster greater understanding of our relationship with water. This awareness is crucial to preserving it and honouring the deep respect it has been accorded through centuries of conscious living, that we seem to have somehow lost in recent times.

The nine months in the womb, where life begins, unfold in water. The act of creation is through the conjunction of waters. Our bodies are composed of 60% or more water, the Earth's surface is covered with 70% of water. By denying our connection with water, the very source of life, we undermine our own sustenance. True healing, beyond external treatments, must restore this fundamental bond and heal the connection with the very source that we have disconnected from. Having severed our connection with this vital life-energy grid, we wander like rats in a maze, fearful of our inevitable death. Perhaps the crisis today is a crisis of deep disconnection and dissonance.

To heal ourselves, and the fractured Earth, we must acknowledge water as a life-giving source. We need to reconnect with the web of connections it forms, like the nerves in our bodies, and the vast network of water channels across the Earth. We must revive the memory of water embedded in our consciousness since the beginning of time, a memory more powerful than the immateriality of the material world that has distracted and overwhelmed our power of thought and reason, of intuition and memory, ensnaring us in *maya,* the web of illusion and delusion.

The rivers in which we immerse ourselves no longer have the power to revive us. Where is the free-flowing Ganga, sought at Kailash, that emerged at Gaumukh and danced her childhood through the Himalaya, reaching maturity in the grandeur and immense power evident at Haridwar, where our ashes have been immersed from lifetime to lifetime? The Ganga unites with the serene Yamuna at Prayagraj, where the subtle presence of the mythical Saraswati reminds us of the wisdom of ages past, now hidden. After Prayagraj, the name of Ganga remains, but the form is of Yamuna. She has given life to millions whom she has blessed in her journey, finally merging with the ocean at Ganga Sagar where Kapil Muni had revealed to the descendants of Maharaja Sagar, that she must be sought for salvation. Dammed and polluted along her entire course by our own hands, who will now offer us salvation? The pollution we see is our sins that she can no longer wash away. And when her tears run dry and there are no sacred waters left on her mined beds, who will hear mankind's grief?